Maple via Calculus

A Tutorial Approach

OTHER MAPLE TITLES

THEORETICAL METHODS IN THE PHYSICAL SCIENCES:
AN INTRODUCTION TO PROBLEM SOLVING USING MAPLE V
William E. Baylis
ISBN 0-8176-3715-X
ISBN 3-7643-3715-X

MAPLE V: MATHEMATICS AND ITS APPLICATION
Proceedings of the Maple Summer Workshop and Symposium, Rensselaer Polytechnic Institute, Troy, New York, August 9-13, 1994
Robert J. Lopez, Editor
ISBN 0-8176-3791-5
ISBN 3-7643-3791-5

MATHEMATICAL COMPUTATION WITH MAPLE V:
IDEAS AND APPLICATION
Proceedings of the Maple Summer Workshop and Symposium, University of Michigan, Ann Arbor, June 28-30, 1993
Thomas Lee, Editor
ISBN 0-8176-3724-9
ISBN 3-7643-3724-9

Robert J. Lopez

Maple via Calculus

A Tutorial Approach

1994

Birkhäuser

Boston • Basel • Berlin

Robert J. Lopez
Department of Mathematics
Rose-Hulman Institute of Technology
5500 Wabash Avenue
Terre Haute, IN 47803

Library of Congress Cataloging-in-Publication Data

Lopez, Robert J., 1941-
Maple via calculus : a tutorial approach / Robert J. Lopez.
p. cm.
Includes bibliographical references and index.
ISBN 0-8176-3771-0 (acid-free)
1. Calculus--Data processing. 2. Maple (Computer file)
I. Title.
QA303.5.D37L66 1994 94-21604
515 '.078--dc20 CIP

Printed on acid-free paper

ISBN 0-8176-3771-0
ISBN 3-7643-3771-0

Text prepared by the author.
Printed and bound by Grafacon, Hudson, MA
Printed in the U.S.A.
9 8 7 6 5 4 3 2 1

TABLE OF CONTENTS

Unit		
1	Basic Plotting	1
2	Parametric Equations	7
3	Optimization Problems	16
4	Interpolation	19
5	Conic Through Five Points	23
6	An Implicit Function	26
7	Inverse Functions	30
8	Partial Fraction Decomposition	34
9	Derivatives by Definition	38
10	Implicit Differentiation	43
11	Taylor Polynomials	44
12	Teaching the Definite Integral	49
13	Deriving Simpson's Rule	55
14	Numerical Integration	57
15	Improper Integrals	62
16	Integration by Trig Substitution	68
17	Integration by Parts	72
18	Integration by Parts Twice	73
19	Surface Area of a Solid of Revolution	76
20	A Separable Differential Equation	80
21	Newton's Law of Cooling	83
22	Logistic Growth	85
23	L'Hôpital's Rule	88
24	Lines and Planes	91
25	Curvature from Every Angle	97
26	The Lagrange Multiplier, Part One	133
27	The Lagrange Multiplier, Part Two	140
28	The Lagrange Multiplier, Part Three	160
29	Iterated Integration	163

PREFACE

Modern software tools like Maple have the potential to alter radically the way mathematics is taught, learned, and done. Bringing such tools into the classroom during lectures, assignments, and examinations means that new ways of looking at mathematics can become permanent fixtures of the curriculum. It is universal access that will make a software-based approach to mathematics become the norm.

In 1988, with NSF funding under an ILI grant, I had the opportunity to bring Maple into the calculus classroom at Rose-Hulman Institute of Technology. Since then a new curriculum based on the availability of computer algebra systems has evolved at RHIT and in my own courses. This volume contains a record of some of the insights gained into pedagogy using Maple in calculus.

The activities and ideas captured in these Maple worksheets reflect concepts in calculus implemented in Maple. There is an overt message to the reader that carries with it a side effect. However, it is possible that for one reader the side effect is the message and the message is the side effect!

I had intended to put before my audience examples extracted from my Maple based curriculum to entice a wider acceptance of the benefits of making a computer algebra system become the basis of a revised calculus syllabus. By examples I had hoped to demonstrate the "rightness" of using software tools for teaching and learning calculus.

As the by-product of these exercises I believed that the reader would learn enough about Maple syntax to become users and creators on their own. But as the volume took shape it became clear that one could just as easily read this work to learn Maple and as a by-product see that Maple in the classroom led to a creative and interesting curriculum.

So, whichever message is most appropriate for the reader is the one that will come across through these pages. Readers who already are familiar with Maple will perhaps see some calculus activities of interest and be inspired to rethink their own syllabus. Readers who seek insight into Maple syntax will find examples of Maple interactions appropriate for a calculus course, and they will perhaps also find that their vision of what should happen during a calculus course has changed.

As these materials were developed and used in the classroom, they became the basis for presentations and workshops in North America and Europe. Audience reactions were positive. Even in those isolated instances where some listeners felt that basing mathematics teaching and learning on a computer algebra system was completely inappropriate I found that the ensuing interactions helped me articulate my thinking more clearly and precisely. I therefore wish to express my appreciation to all my students and to all the members of my various audiences whose feedback, reactions, and suggestions have led to improvements in my work and in this volume.

If this work sparks additional creativity on the part of a single reader the energies expended during its production will have been well spent. My greatest wish then would be that I, in turn, could learn from my readers.

INTRODUCTION

This volume started out as a collection of Maple worksheets generated on a Macintosh computer. In fact, that was the form of the manuscript at the copyedit stage where an excellent copyeditor at Birkhäuser Boston made a number of suggestions for a consistent style. These suggestions were actually implemented in a manuscript that was submitted as the final version.

Alas, it was not to be! Waterloo Maple Software is in the process of developing a suite of authoring tools that provide for the conversion of Maple worksheets into LaTeX. These tools improve on the existing capability in Maple V Release 3. For example, automatic inclusion of graphs into the LaTeX code is possible. Alas, I was asked if I would be a tester of the tools and of the authoring process. And alas, I said yes.

Readers familiar with LaTeX will understand that the look of a manuscript is a function of a template called a style file. Creating a style file is usually an onerous task and the style dictated by the very fine Birkhäuser copyeditor was subjected to the ministrations of an unseen style file. The resulting document approximates the original intentions of author and publisher, and each unit in this volume is as near to the look of a Maple worksheet as could be attained.

Textual narratives are in one font. When I created the narratives I knew what that font was, but only the style file knows for sure what that font turned out to be. But it is distinct, and the reader will have no trouble recognizing a section of text.

Maple input is in a second font, and inputs are preceded by the bullet that only the Macintosh can offer an input prompt. A line of Maple input and the resulting output in mathematical notation generated by Maple therefore look like this:

```
•q := int(sqrt(1 - x^2) , x);
```

$$q := \frac{1}{2}\, x\, \sqrt{1 - x^2} + \frac{1}{2}\, arcsin(x)$$

The spaces in the output are optional and are used to improve readability. The assignment operator (:=) must not contain a space, but otherwise spaces are generally harmless. And an actual worksheet would have the text and the input prompt all left-justified.

A Maple worksheet would not have its graphs enclosed in bounding boxes as we find in this manuscript. Maple on all platforms will give the user the option to separate successive regions (text, input, output) with a solid horizontal line. These separation lines have not been retained by the controlling LaTeX style file.

Within the text regions, references to variables that appear elsewhere as inputs to Maple appear in italics. When entered as input, these variables use the prevailing input font but for contrast, are written in italics in the narrative. Thus, if the text references the expression above, its name will appear as *q* in the text. Names of Maple commands discussed in the narrative are in bold, both to make them stand out and to signal the special nature of such terms.

I take this opportunity to thank the several folks at Waterloo Maple Software who really extended themselves to make this adventure in composition as comfortable and interesting as I could ever have hoped it to be. I look forward to the day when the full suite of authoring tools function on autopilot and require no human intervention. And certainly I need to thank the editors and staff at Birkhäuser for their unprecedented patience in the face of obstacles and deadlines that, like summer storms, pelted down upon us all. Now, though, the harvest is nigh, reap and enjoy.

INDEX OF MAPLE COMMANDS

abs 48, 63
alias 124
allvalues 143
angle 154
array 91, 92, 93
assign 20
assume 32, 33, 62, 66
asympt 82

changevar 68, 166
coeff 35, 147
collect 86, 124
completesquare 148
contourplot 133, 134, 140
convert/list 158
convert/parfrac 34, 37
convert/polynom 44, 46, 90
convert/radical 14
crossprod 92, 93

D 99
denom 34, 35, 36, 89
diff 39, 40, 41, 43, 45, 54, 58, 78, 79, 80, 82, 85, 89, 97, 98, 99, 100, 101, 102, 103, 104, 110, 111, 112, 113, 117, 118, 120, 122, 124, 127, 136, 137, 139,142, 150, 151, 160
Digits 6
display 77, 97, 108, 121, 125, 130, 134, 135, 141, 142, 156
display/insequence=true 10, 18, 108, 126
do 8, 19, 23, 36, 42, 45, 46, 48, 55, 59, 81, 111, 115, 134, 141
dotprod 94
Doubleint 166
dsolve 80, 85

equation 127
evalf 4, 5, 6, 29, 47, 54, 57, 58, 60, 66, 79, 87, 144, 150, 151, 158
evalm 93, 138, 157, 158
expand 5, 15, 106, 110, 147, 148, 162
expand/radical 99

factor 35, 74, 75, 110
fsolve 6, 145, 150, 154

geometry 127
grad 137, 138, 139, 154, 155

implicitplot 24, 27, 108, 134, 141, 142
int 52, 56, 57, 60, 62, 63, 66, 69, 79, 98, 166
Int 60, 63, 64, 65, 68, 71, 73, 97, 98, 99, 166
int/CauchyPrincipalValue 62
intparts 71, 73, 74
isolate 74, 75, 160, 162

leftbox 49, 50, 51
leftsum 51, 52, 53
lhs 118, 138
limit 39, 41, 52, 53, 62, 63, 64, 82, 109
Limit 53, 63, 166
linalg 91, 137, 154, 164
line 127

map 158, 162
matrix 164
matrixplot 164

nops 28, 141
norm 157
normal 35, 118, 145, 153
numer 34, 35, 37, 89, 90, 118, 145, 154

op 29, 138, 155, 158

plot 1, 2, 7, 9, 11, 12, 13, 15, 16, 17, 18, 21, 25, 26, 27, 28, 30, 31, 32, 38, 40, 41, 42, 44, 45, 47, 58, 64, 65, 76, 82, 87, 88, 89, 97, 104, 108, 111, 112, 115, 120, 121, 122, 123, 125, 126, 128, 129, 130, 136, 145, 150, 152, 153, 154, 155, 156, 158
plots 10, 18, 24, 27, 76, 97, 133, 140, 164
point 127
powsubs 147
print 8, 111
product 37

quo 34

readlib 11
rhs 85, 100, 112, 118, 138
rightbox 50
rightsum 51
RootOf 14, 143

scalarmul 93
seq 9, 10, 18, 108, 115, 119, 122, 126, 139, 158
series 90

simplify 5, 11, 12, 28, 32, 33, 38, 56, 68, 69, 74, 75, 82, 86, 101, 102, 104, 106, 113, 118, 120, 162, 163
simplify/power/symbolic 99
simplify/symbolic 69, 101, 102, 110
slope 77, 137
solve 3, 10, 13, 14, 20, 23, 24, 25, 26, 29, 31, 36, 43, 55, 59, 69, 78, 80, 82, 84, 86, 92, 94, 95, 98, 101, 103, 104, 105, 107, 108, 109, 111, 118, 122, 123, 128, 136, 137, 144, 145, 146, 147, 148, 149,150, 151, 152, 155, 157, 158, 163, 165
sort 56
student 49, 57, 68, 71, 73, 77, 137, 147, 160, 166
subs 5, 6, 8, 11, 13, 14, 17, 19, 22, 23, 24, 28, 32, 33, 36, 38, 39, 40, 41, 42, 43, 45, 46, 47, 48, 53, 55, 58, 63, 69, 80, 81, 86, 87, 88, 94, 95, 98, 99, 100, 101, 102, 103, 104, 105, 106, 107, 108, 109, 111, 118, 122, 123, 128, 136, 137, 144, 145, 146, 147, 148, 149, 150, 151, 152, 155, 157, 158, 163, 165
sum 35
Sum 53, 165
surd 11, 12, 13

taylor 44, 46, 90
textplot 76, 97, 121, 125, 130, 156
trapezoid 57, 59

unapply 65, 119, 122
union 87, 106

value 51, 52, 53, 57, 59, 63, 64, 69, 71, 72, 166
vector 158

with(geometry) 127
with(linalg) 91, 137, 154, 164
with(plots) 10, 18, 24, 27, 76, 97, 133, 140, 164
with(student) 49, 57, 68, 71, 73, 77, 147, 160, 166

-> 9, 18, 108, 114, 126
@ 162
$ 45

Unit 1: Basic Plotting

To explore the simplest of Maple's capabilities, let's look at some basic plotting. To begin, enter an expression and tag it with the name f. The assignment operator is the := operator. Please note that you have not created a function in the mathematical sense. The notation $f(2)$ will not be recognized by Maple with the assignment made here.

It is possible to create *functions* but my classroom experience convinces me that the simplest thing for students to start with is *named expressions*, as illustrated in the first Maple input below. Later units in this text will use Maple's functional capabilities where appropriate.

Finally, issue the Maple **plot** command to create a Maple graph.

- ```
  f := x^2;
  ```

$$f := x^2$$

- ```
  plot(f);
  Warning in iris-plot: empty plot
  ```

No domain was given. Hence, no graph. The correct syntax for the **plot** command:

- ```
  plot(f, x = -3..3);
  ```

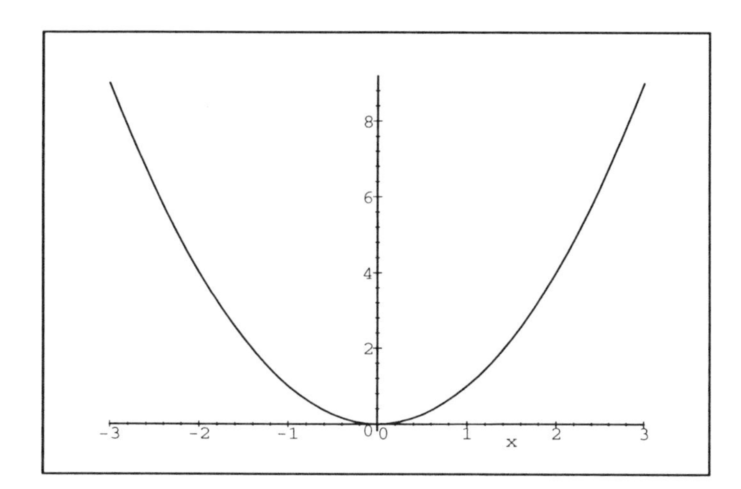

To gain control over the vertical scale, use a second range in the plot command. Maple interprets the first range as the horizontal scale and the second range as the vertical.
  ```

- `plot(f, x = -3..3, y = -1..15);`

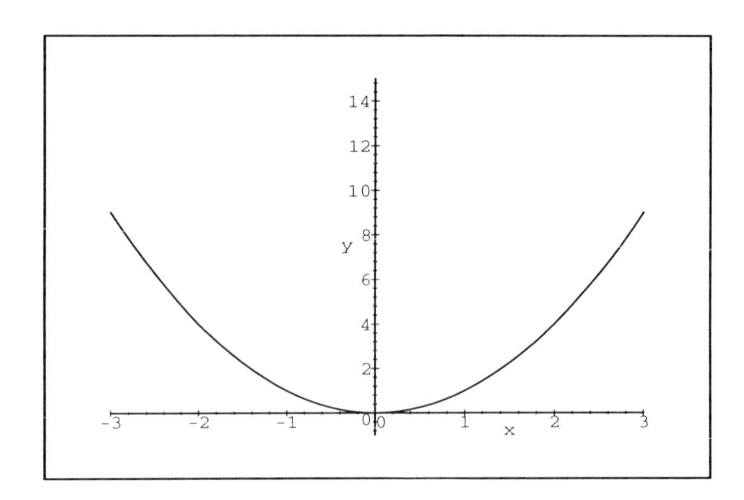

Finally, to gain even greater control over the shape of the graph, plot the function parametrically. In addition to the domain chosen for the parameter, you can also impose a "view window" by using both a horizontal and a vertical range in Maple.

- `plot([x, f, x = -3..3], x = -8..8, y = -1..15);`

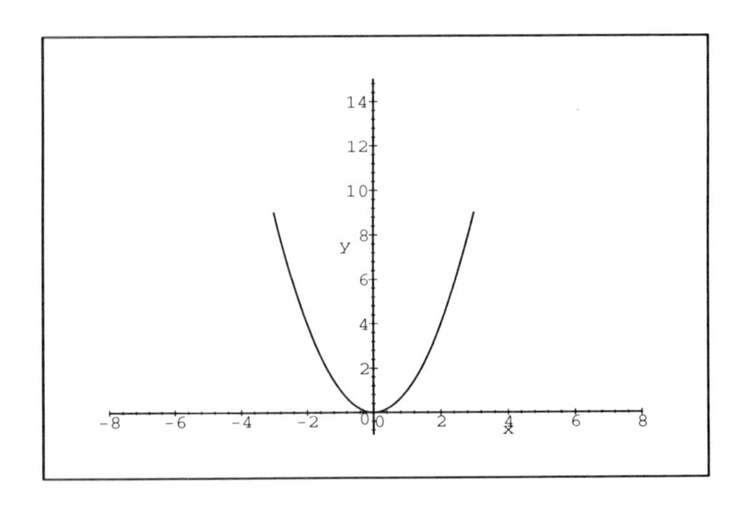

Now that we can plot, what are we going to do with this new power? Suppose we examine the intersection of some simple curves and calculate the points of intersection. Enter a second "function" g.

- `g := 7 + 3*x - 5*x^2;`

$$g := 7 + 3\,x - 5\,x^2$$

Plot both functions on one set of axes to see what to expect for intersections.

- `plot({f, g}, x = -3..3);`

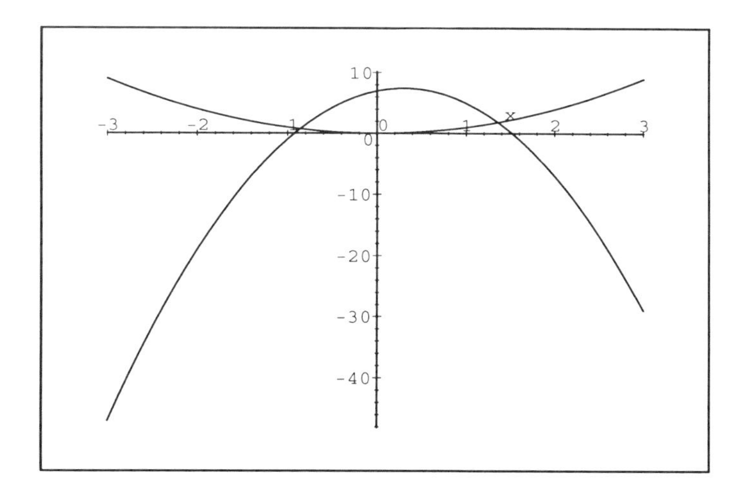

There are two intersection points to compute, and the x-coordinates of each point has magnitude of about one. To compute the points exactly, we can set f and g equal and solve for x.

- `solve(f = g, x);`

$$\frac{1}{4}+\frac{1}{12}\sqrt{177}, \frac{1}{4}-\frac{1}{12}\sqrt{177}$$

The solve command has returned an *expression sequence* containing the two roots, listed in sequence. The ease with which these solutions can be referenced and used is an essential feature of a computer algebra system.

In order to understand the options available, we stop to clarify three Maple constructs:

Expression Sequence:	a, b, c
List:	$[a, b, c]$
Set:	$\{a, b, c\}$

The *expression sequence* is the content of both the list and the set.

The *list* retains order and multiplicity. The lists $[a, b, c]$ and $[a, a, b, c]$ are distinct, as are the lists $[a, b, c]$ and $[a, c, b]$.

The *set* is mathematically faithful. The sets $\{a, b, c\}$ and $\{a, a, b, c\}$ are the same set; so are the sets $\{a, b, c\}$ and $\{a, c, b\}$.

To reference an individual member of any of these three data structures, it is convenient to tag the data structure with a referencing name and to use "selector brackets." Thus,

$$\begin{array}{lcl}
\mathbf{q1 := a, b, c} & \Longrightarrow & q1[2] \text{ is } b\text{, etc.} \\
\mathbf{q2 := [a, b, c]} & \Longrightarrow & q2[2] \text{ is } b\text{, etc.} \\
\mathbf{q3 := \{a, b, c\}} & \Longrightarrow & q3[2] \text{ is } b\text{, etc.}
\end{array}$$

One final word is needed here. Maple will not allow the expression sequence as an argument to most commands, but it will usually allow sets and lists. We illustrate this distinction with the **evalf** command, which converts symbolic representations of numbers to floating point form.

To tag the output of the **solve** command, issue the command as follows:

- `q := solve(f = g, x);`

$$q := \frac{1}{4} + \frac{1}{12}\sqrt{177}, \frac{1}{4} - \frac{1}{12}\sqrt{177}$$

Just for practice, reference the individual members of the expression sequence q.

- `q[1];`

$$\frac{1}{4} + \frac{1}{12}\sqrt{177}$$

- `q[2];`

$$\frac{1}{4} - \frac{1}{12}\sqrt{177}$$

Either of these roots can be individually submitted to the **evalf** command.

- `evalf(q[1]);`

$$1.358677892$$

- `evalf(q[2]);`

$$-.8586778920$$

In Release 3, you can now apply the **evalf** command to an expression sequence *if it has a name*. This freedom does not extend to other Maple commands.

- `evalf(q);`

$$1.358677892, -.8586778920$$

If the expression sequence q is the "inside" of either a list or a set, the **evalf** command will return an equivalent data structure of floating point numbers.

- `evalf([q]);`

$$[1.358677892, -.8586778920]$$

- `evalf({g});`

$$1.358677892, -.8586778920$$

In Release 3, it will be most convenient to work with the expression sequence of roots, *provided this expression sequence has a name.*

- `g1 := evalf(g);`

$$g1 := 1.358677892, -.8586778920$$

- `q1[1];`

$$1.358677892$$

- `q1[2];`

$$-.8586778920$$

Finally, let's see how we would obtain the y-coordinates corresponding to the two x-coordinates just calculated. Note that mathematically we would "plug in" or substitute the x-value into the appropriate expression.

- `y1 := subs(x = q[1], f);`

$$y1 := \left(\frac{1}{4} + \frac{1}{12}\sqrt{177}\right)^2$$

- `y2 := subs(x = q[2], f);`

$$y2 := \left(\frac{1}{4} - \frac{1}{12}\sqrt{177}\right)^2$$

Obviously, there is a need for some simplification. Two options come to mind. We can try to *expand* the expressions for $y1$ and $y2$, or we can try to *simplify* these expressions.

- `simplify(y1);`

$$\frac{1}{144}\left(3 + \sqrt{177}\right)^2$$

- `expand(y1);`

$$\frac{31}{24} + \frac{1}{24}\sqrt{177}$$

The more natural action is to *expand* the term being squared. Hence, the **expand** command produced the more appropriate result.

Sometimes it is necessary to work completely numerically. Consider the following:

- `yy1 := subs(x = q[1], f);`

$$yy1 := 1.8460056121$$

Finally, let's examine a few more features of numerical work in Maple. Begin by obtaining more digits in the floating-point versions of the x-coordinates at the intersections.

- `xx1 := evalf(q[1], 30);`

$$xx1 := 1.35867789130417256042050256784$$

- `subs(x = xx1, f);`

$$1.846005612$$

Note that Maple did not transfer to the calculation of y the need for more digits. One way to dictate that outcome is via the **evalf** command.

- `evalf(subs(x = q[1], f), 30);`

$$1.84600561231875294687691795058$$

A second way to generate more digits is to make a global in the number of digits. Use the system variable **Digits** , paying attention to the capitalization of the first letter.

- `Digits := 30;`

$$Digits := 30$$

- `evalf(q[1]);`

$$1.35867789130417256042050256784$$

- `evalf(subs(x=q[1],f));`

$$1.84600561231875294687691795058$$

The default value of **Digits** is 10. To restore the default value, set **Digits** back to 10.

- `Digits := 10;`

$$Digits := 10$$

The last feature of solving equations we should discuss is the direct numeric calculation of roots of equations. This requires that we look at the **fsolve** command.

- `fsolve(f = g, x);`

$$-.8586778913, 1.358677891$$

In this instance, **fsolve** delivered both solutions. In the event that **fsolve** should fail to deliver all solutions or fail to deliver the desired solution, there is an option for telling **fsolve** where to look for the desired solution.

- `fsolve(f = g, x, 0..2);`

$$1.35867789130417256042050256784$$

Unit 2: Parametric Equations

Parametric plotting is so easy with a computer algebra system that it can be, and is often, introduced very early in a calculus course that is based on software tools. Hence, it is appropriate to consider how a beginning calculus student might be helped by Maple when learning about parametric representations of curves.

A first device would be the creation of a numeric table, followed by the direct plotting of the data from this table. A second aproach would be analytic, with the parametric representation being converted to a cartesian representation and the curve plotted in cartesian coordinates. The interesting issues of domain equivalence usually arise in this context.

Let's illustrate these strategies with an example, an illustration that will demonstrate viable strategies for naming and using Maple variables.

For simplicity, let's take $x(t) = 1 + t^3$ and $y(t) = 1 - t^2$. Note the use of the names xt and yt, rather than $x(t)$ and $y(t)$, names that would lead to great confusion in Maple.

- `xt := 1 + t^3;`

$$xt := 1 + t^3$$

- `yt := 1 - t^2;`

$$yt := 1 - t^2$$

Obtaining a graph is certainly a useful first step.

- `plot([xt, yt, t = -2..2]);`

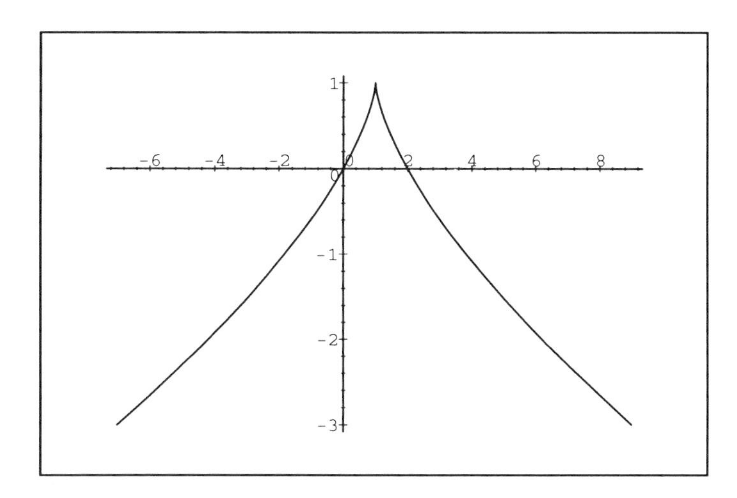

This graph hides the parameter dependence of the points (x, y). From this plot we cannot determine the "direction of increasing values" of the parameter along the curve. One device to rectify that loss would be numeric. Hence, a table of values would indicate which points correspond to $t = -2$ and $t = 2$. We construct such a table with a Maple loop.

- ```
  for k from 0 to 10 do
    t.k := -2 + (4/10)*k;
    P.k := subs(t=t.k, [xt, yt]);
    print(t.k, P.k);
  od:
  ```

$$-2, [-7, -3]$$

$$\frac{-8}{5}, \left[\frac{-387}{125}, \frac{-39}{25}\right]$$

$$\frac{-6}{5}, \left[\frac{-91}{125}, \frac{-11}{25}\right]$$

$$\frac{-4}{5}, \left[\frac{61}{125}, \frac{9}{25}\right]$$

$$\frac{-2}{5}, \left[\frac{117}{125}, \frac{21}{25}\right]$$

$$0, [1, 1]$$

$$\frac{2}{5}, \left[\frac{133}{125}, \frac{21}{25}\right]$$

$$\frac{4}{5}, \left[\frac{189}{125}, \frac{9}{25}\right]$$

$$\frac{6}{5}, \left[\frac{341}{125}, \frac{-11}{25}\right]$$

$$\frac{8}{5}, \left[\frac{637}{125}, \frac{-39}{25}\right]$$

$$2, [9, -3]$$

The normal line termination in Maple is the semicolon. The colon as a line terminator is a *silencer*, suppressing output from that command. The punctuation on the loop terminator **od** determines the output from the whole loop. To suppress the additional printing of the individual values of t, we used the colon on the loop itself. Then, in order to have output from the loop, we used a **print** statement.

Another interesting syntactic issue is *concatenation*, the formation of new (variable) names from old ones. In this loop the "old" names are "t" and the numbers 0, 1,..., 10. The symbol "t.k" dynamically creates the new variables $t0, t1, \cdots, t10$, and can be thought of as a subscripting operation.

Additionally, we have represented points on the curve as the lists $P0, P1, \cdots, P10$, which are syntactically correct for use in plotting. In fact, we can create the desired parametric plot by connecting these points with line segments.
  ```

- `plot([seq(P.k, k = 0..10)]);`

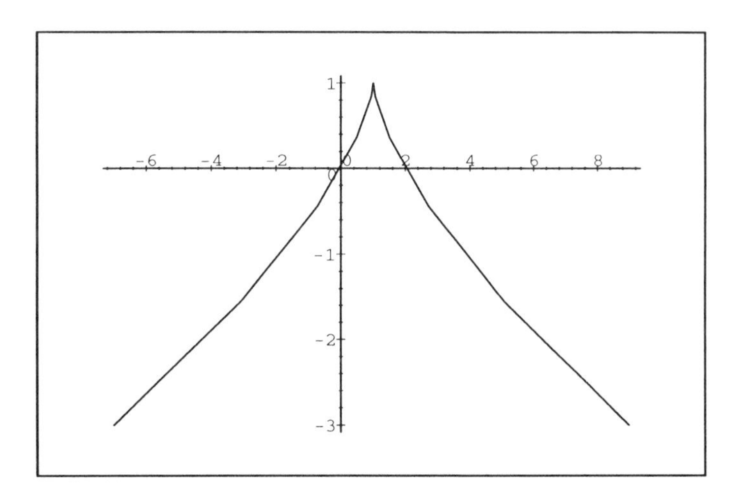

Except for some problems with resolution, we have again obtained the plot of the given curve. Note the use of Maple's sequence operator, **seq** with which we have created the appropriate sequence of points sent to the **plot** command.

While we can now determine the direction of the tracing of the parametrically given curve, it might be useful to describe a method for animating this trace. Animation requires that a series of snapshots be created and displayed in succession. Each snapshot should show a bit more of the tracing than the previous one.

One way to do this would be by creating a function $f(z)$, z in $[-12, 2]$, which returns a plot of the parametric curve for t in the interval $[-2, z]$. Hence, as z progresses through the interval $[-2, 2]$, the values of $f(z)$ would be the desired snapshots of the tracings of the parametric curve.

- `f := z -> plot([xt, yt, t = -2..z]);`

$$f := z \to \mathrm{plot}(\,[\,xt, yt, t = -2..z\,]\,)$$

```
f(1.5);
```

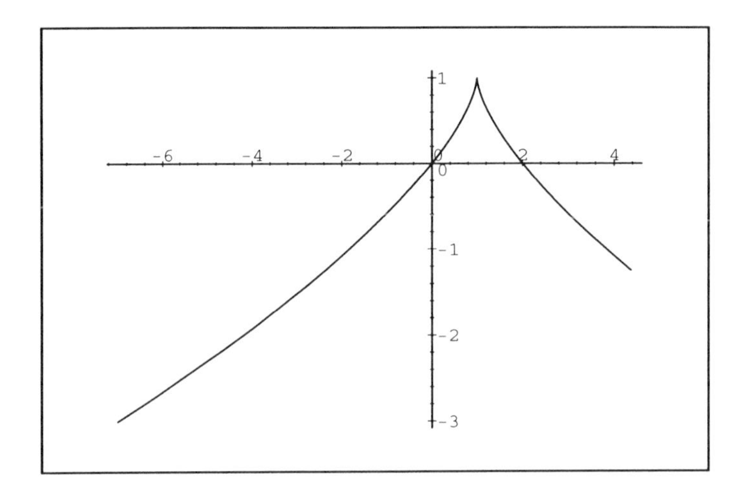

The function $f(z)$ indeed returns a graph of the parametric curve for t in the interval $[-2, z]$. We now need to create a sequence of these snapshots and display the resulting "movie." We'll use Maple's sequence operator, **seq** to create the sequence of snapshots, and we'll use Maple's **display** command (found in the plots package) to create the animation. We already have a sequence of uniformly spaced values $(t0, t1, ..., t10)$ in the interval $[-2, 2]$.

```
with(plots):
display([seq(f(t.k), k=0..10)], insequence=true);
```

The only way to appreciate animation is to watch it in progress. It cannot be adequately represented statically on paper. Syntactically, however, notice the use of the **display** command, which pastes together individual plots except when the **insequence** flag is set to **true**. Then, instead of putting all the plots onto one set of axes, they are shown in sequence as in a movie.

Finally, we examine the analytic task of eliminating the parameter t and obtaining a cartesian representation of this curve. We might, for example, solve $x(t)$ for $t(x)$ and plug that into $y(t)$. Or we might solve $y(t)$ for $t(y)$ and plug that into $x(t)$. Clearly, the success of such operations will depend on the complexity of the particular expressions involved. What will always be the case, however, is the care we take to avoid using as a name (on the left of the assignment operator :=) any working variable we are using on the right of the assignment operator.

```
T := solve(xt = x, t);
```

$$T := (-1+x)^{1/3}, -\frac{1}{2}(-1+x)^{1/3} + \frac{1}{2} I\sqrt{3}(-1+x)^{1/3}, \\ -\frac{1}{2}(-1+x)^{1/3} - \frac{1}{2} I\sqrt{3}(-1+x)^{1/3}$$

We solved for t in the equation $x = 1+t^3$. It was with great foresight that we used xt for the name of the expression in $x(t)$. By not using x as the name of this expression we were then free to use the letter x as a variable in the equation $x = 1 + t^3$. Likewise, we are careful not to call the output of the solve command t . Rather, we have used T , a distinct name. Such care avoids syntax errors and infinite recursions (caused by assigning a name to itself).

We want just the first value for t, the real one.

- `Y := subs(t = T[1], yt);`

$$Y := 1 - (-1 + x)^{2/3}$$

Again, we avoid using as a name (on the left) what might later be needed as a variable on the right. The letter y is free at the moment, but if we redo this calculation by solving $y(t) = y$ for $t(y)$, we will probably want y to be a free variable. Hence, we have used Y.

- `plot(Y, x = -6..8);`

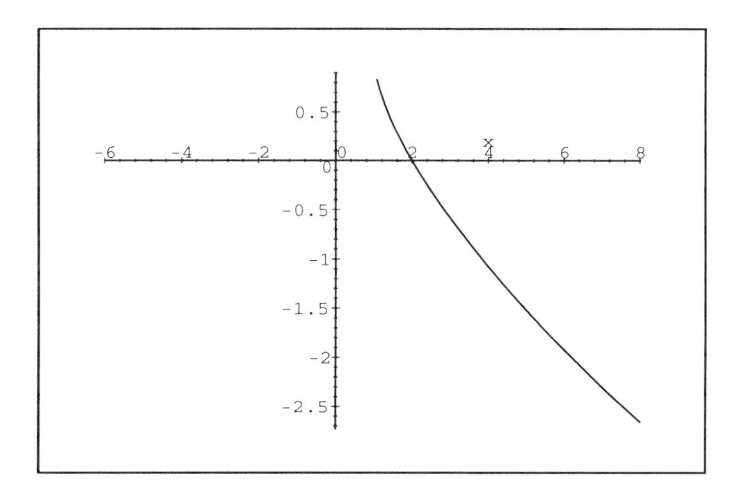

Oh my, what has happened? The cartesian representation does not resemble the parametric plot. Is this because of something inherent in the cartesian representation, or is it attributable to some other phenomenon?

To make a long story short the difficulty we have uncovered is that in Release 3, Maple opts to return a complex number for $x^{1/3}$. This complex number is the principal branch of the cube root function, correct for complex arithmetic but very inappropriate for freshman calculus.

Release 3 does, however, contain a new function that will return the real branch of the cube root function. This function, **surd** , is illustrated below in a simpler context.

- `simplify((-8)^(1/3));`

$$1 + I\sqrt{3}$$

This choice of branch, instead of the choice -2, will be objectionable in most freshman calculus classes in which it is discovered. However,

- `readlib(surd):`

```
simplify(surd(-8,3));
```

$$-2$$

In fact, the following plot "fails" because Maple generates complex numbers for $x < 0$.

- `plot(x^(1/3), x = -8..8);`

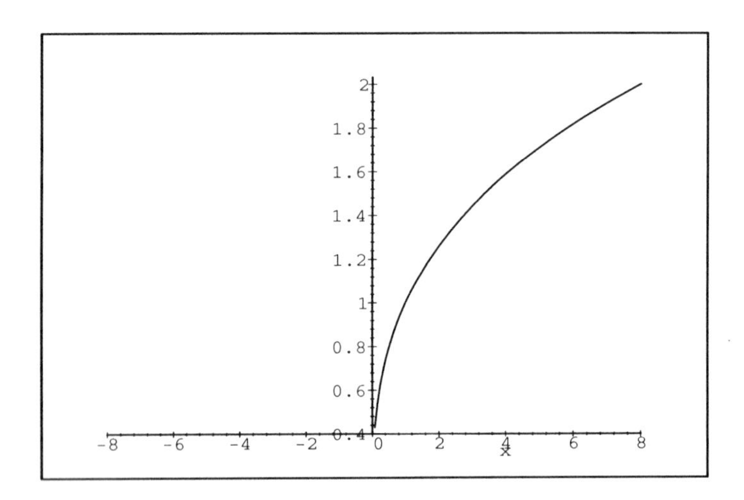

Compare this behavior to

- `plot(surd(x,3), x = -8..8);`

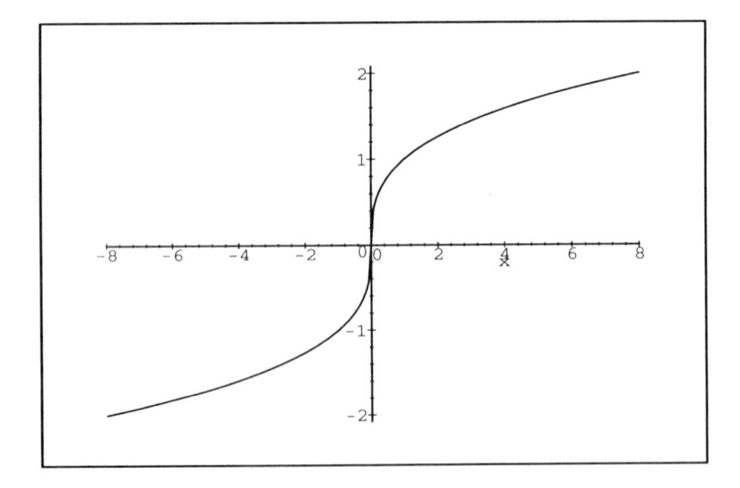

Thus, if we express our cartesian version of the parametrically given curve using the **surd** function, we should see a curve that resembles the graph we obtained parametrically.

- `plot(1-surd(1-x,3)^2, x = -6..8);`

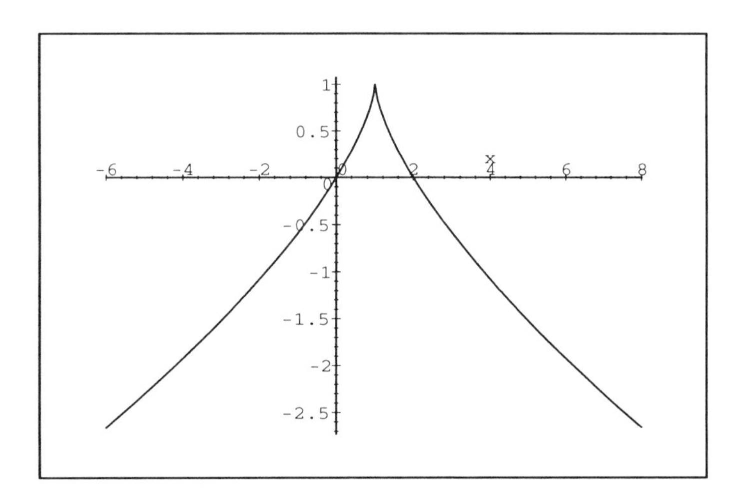

The **surd** function gives the expected real roots, but we really hope that students don't create the mnemonic "absurd" for this functionality.

We complete this discussion by computing the cartesian representation by solving the equation $y = 1 - t^2$ for t and then substituting $t(y)$ into $x(t)$.

- `ty := solve(yt = y, t);`

$$ty := -\sqrt{1-y}, \sqrt{1-y}$$

Having used T above, we here use ty for $t(y)$. If we knew we were not going to execute parts of this worksheet again, we could overwrite T by using T instead of ty. However, if we did that, any attempt to recalculate a portion of the worksheet that contained the now ambiguous T would be difficult. Hence, we stick to the strategy of using distinct names in the worksheet whenever we would risk creating ambiguity while experimenting with portions of an already calculated session.

The function $t(y)$ has two branches, the first of which gives t values that are non-positive, and the second of which gives t values that are nonnegative. Since t was assigned both negative and positive values for the parametric representation of the curve, it is not immediately obvious that we can ignore one of these branches.

- `x1 := subs(t = ty[1], xt);`

$$x1 := 1 - (1-y)^{3/2}$$

- `x2 := subs(t = ty[2], xt);`

$$x2 := 1 + (1-y)^{3/2}$$

Solve each of these for $y = y(x)$.

- `y1 := solve(x1 = x, y);`

$$y1 := 1 - (1-x)^{2/3}, 1 - \left(-\frac{1}{2}(1-x)^{1/3} - \frac{1}{2} I\sqrt{3}(1-x)^{1/3}\right)^2,$$
$$1 - \left(-\frac{1}{2}(1-x)^{1/3} + \frac{1}{2} I\sqrt{3}(1-x)^{1/3}\right)^2$$

- `y2 := solve(x2 = x, y);`

$$y2 := 1 - (-1+x)^{2/3}, 1 - \left(-\frac{1}{2}(-1+x)^{1/3} + \frac{1}{2} I\sqrt{3}(-1+x)^{1/3}\right)^2,$$
$$1 - \left(-\frac{1}{2}(-1+x)^{1/3} - \frac{1}{2} I\sqrt{3}(-1+x)^{1/3}\right)^2$$

The first member in the sequence $y2$ agrees with our earlier result. The first member of $y1$, because of the square on $(1-x)$, is actually the same as the first member of $y2$. Hence, the cartesian representation is "unique."

There is another, more direct approach to converting parametric equations to cartesian form. This consists of using Maple's **solve** command to solve simultaneously the equations $xt = x$, and $yt = y$ for y and t. We will get back $y(x)$ and $t(x)$, from which we discard $t(x)$ to obtain the direct solution for $y(x)$.

Thus,

- `q := solve({xt = x, yt = y}, {y, t});`

$$q := \left\{y = 1 - \mathrm{RootOf}(1 + \_Z^3 - x)^2, t = \mathrm{RootOf}(1 + \_Z^3 - x)\right\}$$

Before we deal with the **RootOf** structure, we'll extract the result for y since it is the only part of the solution we really want.

- `Y := subs(q, y);`

$$Y := 1 - \mathrm{RootOf}(1 + \_Z^3 - x)^2$$

The cure for a **RootOf** is either the **allvalues** command or a conversion to radical form. We try the latter.

- `convert(Y, radical);`

$$1 - (-1+x)^{2/3}$$

We have recovered the cartesian form of $y(x)$ that was so painfully extracted earlier. However, the effort it took to obtain $y(x)$ by first finding $t = t(x)$ showed a conceptual basis for how to think about the conversion from parametric to cartesian forms.

Unit 3: Optimization Problems

The hardest part of solving an optimization problem is formulating the objective function for it. To deal with this difficulty you might consider assigning optimization problems early in the calculus course and continuing them every assignment for the whole term. To "solve" an optimization problem before the derivative is formally available, plot the objective function and use the graph to extract an approximation to the extreme value.

Here is the classic example of the "Box Problem," wherein equal squares are cut from the corners of a sheet of cardboard of dimension 8.5 by 11 inches, the resulting flaps folded, and the volume of the box so formed is maximized.

- `v := x*(17/2 - 2*x)*(11 - 2*x);`

$$v := x\left(\frac{17}{2} - 2\,x\right)(11 - 2\,x)$$

While it is not essential that we expand this factored form of the expression for the volume, the way to get Maple to clear the parentheses is

- `expand(v);`

$$\frac{187}{2}\,x - 39\,x^2 + 4\,x^3$$

However, we are really interested in the graph of this expression as a function of x , the length of the edge of the square corner removed.

- `plot(v, x = 0..7);`

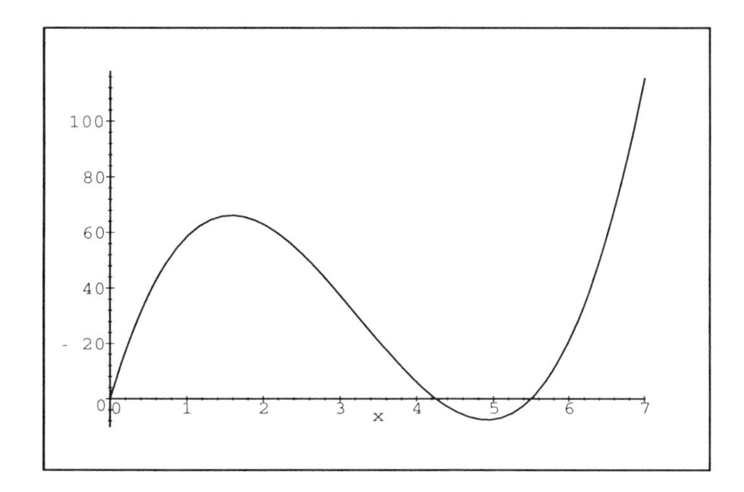

An examination of the graph indicates that there are values of the independent variable x for which the volume of the box becomes negative. Clearly, this happens when the corners removed "overlap." In addition, Release 3 plots have a digitizer; clicking the mouse button generates the coordinates of the point at the tip of the mouse arrow. This requires that the user place the mouse arrow at the location of the point whose coordinates are desired.

An estimate of the coordinates of the relative maximum for the volume would therefore be $x = 1.613, v = 66.29$.

Some early versions of the Maple interface did not have a digitizer available for plots. Where this is not available, estimates of the extreme value could be improved by replotting with altered scales.

- `plot({v, 70}, x = 0..4);`

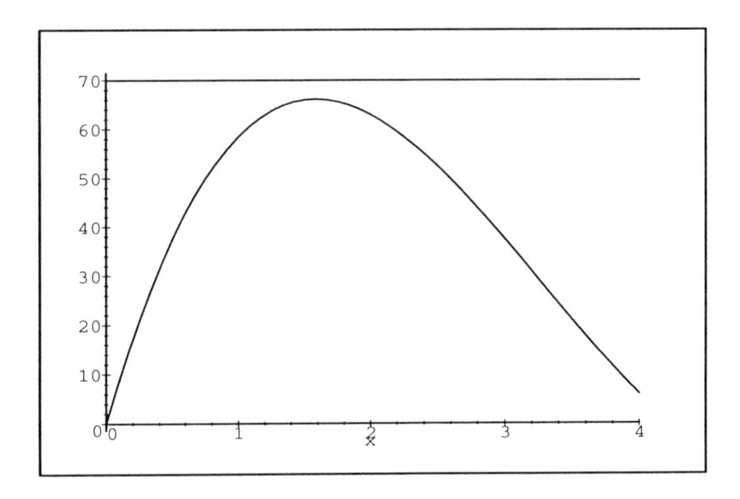

- `plot({v, 66}, x = 1..2);`

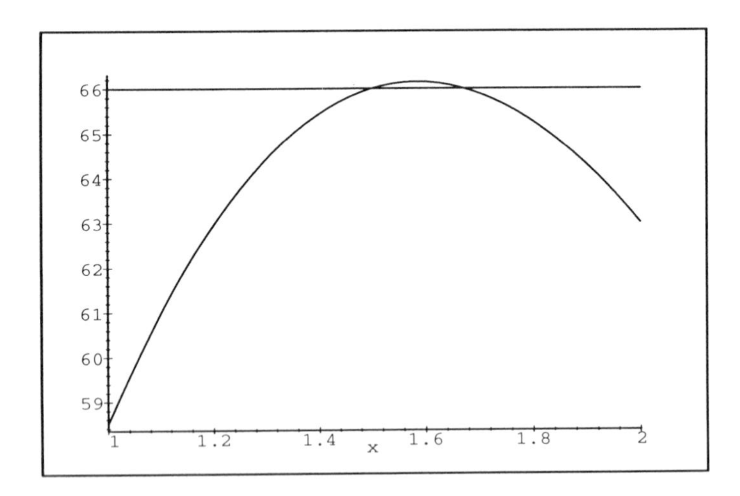

And so on.

A second example of an optimization problem that profits from technology is the minimization of the distance from a point to a curve. Obtaining a graph of the appropriate objective function helps make concrete what is happening with such problems. Thus, for example, we find the function d that expresses the distance from the point $(3, 5)$ to the graph of $y = sin(x)$.

- `d := sqrt( (x-3)^2 + (5-sin(x))^2 );`

$$d := \sqrt{(x-3)^2 + (5-\sin(x))^2}$$

- `plot(d, x = 0..2*Pi,'d' = -1..7);`

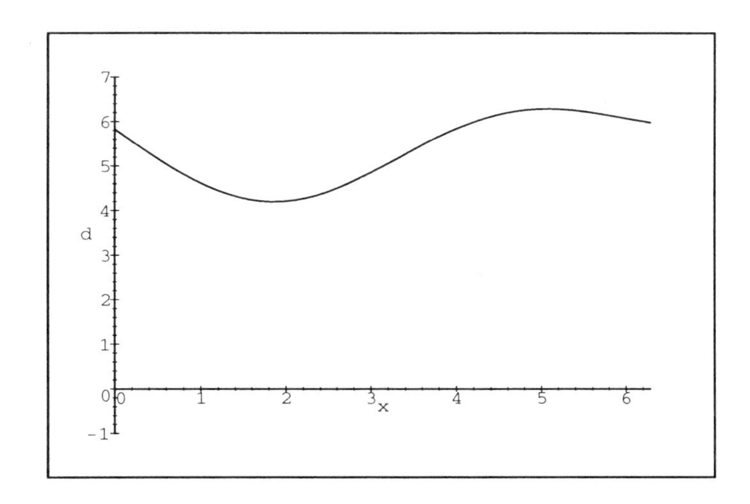

In the **plot** command, the range on the independent variable x requires an equation. For the second range, the one for the dependent variable d, an equation is optional. Whatever letter does appear in this equation will be written adjacent to the vertical axis. Since the variable d already has a value, the quotes around the letter d in the **plot** command prevent its assigned *contents* from being written on the axis.

From the graph we can see that $(3, 5)$ is closest to a point near $(2, sin(2))$. It helps if we draw lines from the point $(3, 5)$ to points on the curve $y = sin(x)$. For this, a general form of a line from $(3, 5)$ to a point on $y = sin(x)$ is very useful.

- `Y := (5-sin(a))/(3-a)*(x-a) + sin(a);`

$$Y := \frac{(5-\sin(a))(x-a)}{3-a} + \sin(a)$$

Three specific realizations of this line are given by

- `Y1 := subs(a = 1, Y);`

$$Y1 := \frac{1}{2}(5-\sin(1))(x-1) + \sin(1)$$

- `Y2 := subs(a = 2, Y);`

$$Y2 := (5-\sin(2))(x-2) + \sin(2)$$

- `Y3 := subs(a = 5/2, Y);`

$$Y3 := 2\left(5-\sin\left(\frac{5}{2}\right)\right)\left(x-\frac{5}{2}\right) + \sin\left(\frac{5}{2}\right)$$

A graph of these lines along with a graph of $sin(x)$ can be obtained parametrically by

```
• plot({[x, sin(x), x = 0..2*Pi], [x, Y1, x = 1..3   ],
        [x, Y2,     x = 2..3   ], [x, Y3, x = 5/2..3 ]});
```

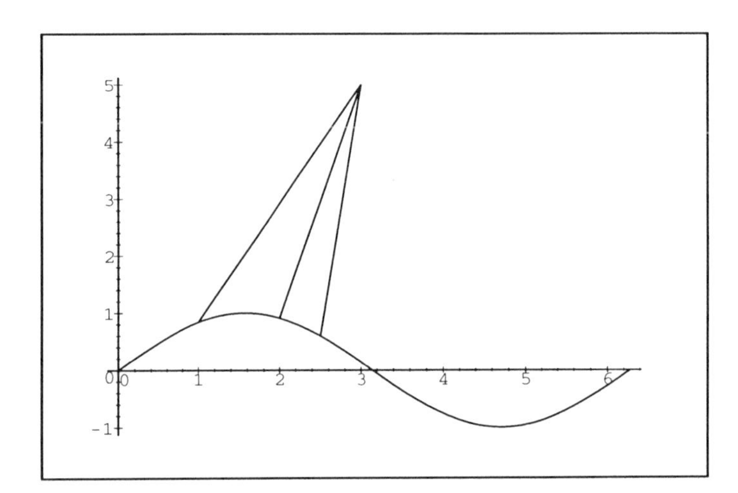

Readers with great insight will probably see that here is a valid opportunity to make use of animation. Suppose we create an animation in which the line measuring the distance from the point $(3, 5)$ to the graph of $sin(x)$ sweeps across $sin(x)$. What would be an efficient way to do this?

For starters, note that the simplest way to have drawn the lines from $(3, 5)$ to $sin(x)$ would have been to use Maple's "connect-the-dots"" approach to plotting lines. We needed only to have given to the **plot** command the endpoints of the line for Maple to have drawn the connecting segment. The approach we took above is not unwarranted early in a calculus course, however. The ingredients of this animation will be

1. a sequence of uniformly spaced x-coordinates at which rays from $(3, 5)$ will hit the graph of $sin(x)$.

2. a function whose input is one of these uniformly spaced x-coordinates, and whose output is the plot data structure of one frame of the animation. This one frame contains an image of the graph of $sin(x)$ and the image of one ray from $(3, 5)$.

3. a call to the Maple function that will join and "show" the frames of the movie. Since this functionality is in the plots package, we need to begin by accessing that package.

Thus:

```
• with(plots):
  A := [seq(k/4, k=0..24)]:
  f := a -> plot({[x, sin(x), x = 0..2*Pi], [[3,5], [a,sin(a)]]}):
  display([seq(f(a), a = A)], insequence=true);
```

Clearly, paper is not the medium from which to view an animation. But feel free to execute the above code and to experience the resulting movie.

Unit 4: Interpolation

Having practiced the analysis of common functions and their graphs, let's examine the task of synthesis. Let's fit curves to given sets of points in the sense of interpolation, passing the curves through each prescribed point.

We begin by fitting the parabola

$$y = ax^2 + bx + c \tag{1}$$

to the points $(2,3), (-3,5), (-4,-1)$.

We will use this example to demonstrate the use of loops and concatenation in Maple. Since this interpolation task is equivalent to setting up three equations in the three unknowns a, b, and c, we anticipate the outcome and begin by entering the parabola y and two lists. The first list is the list of x-coordinates, and the second list is the list of corresponding y-coordinates.

- `y := a*x^2 + b*x + c;`

$$y := a\,x^2 + b\,x + c$$

- `X := [2, -3, -4];`

$$X := [2, -3, -4]$$

- `Y := [3, 5, -1];`

$$Y := [3, 5, -1]$$

Notice the use of uppercase letters for X and Y, the names of the lists just created.

Next, we need to substitute each of the x-coordinates into the expression for y and set the result equal to the corresponding y-coordinate for that point.

For the first point, the x- and y-coordinates can be extracted from the lists X and Y by the Maple notation $X[1]$ and $Y[1]$. Since this is to be repeated for each of the three points, we need a looping structure to handle such a repetitive chore.

- `for k from 1 to 3 do e.k := subs(x = X[k], y) = Y[k]; od;`

$$e1 := 4\,a + 2\,b + c = 3$$

$$e2 := 9\,a - 3\,b + c = 5$$

$$e3 := 16\,a - 4\,b + c = -1$$

The concatenation operator is the dot. It concatenates, or combines, the name e with 1, e with 2, and e with 3 to form the names $e1$, $e2$, $e3$. Think of these symbols as subscripted variables.

Each equation has been given a separate name: $e1$, $e2$, $e3$. Note the use of the "=" at the right end of the equation to create an equation data-structure in Maple.

Finally, the loop terminator is **od** (which is **do** spelled backwards). These are reserved words, as are **if** and its companion, **fi** .

Now let's solve these three equations for a, b, and c.

```
• q := solve({e1,e2,e3}, {a,b,c});
```

$$q := \left\{b = \frac{-22}{15}, a = \frac{-16}{15}, c = \frac{51}{5}\right\}$$

The **solve** command returns the set q containing equations for the values of a, b, and c which cause parabola (1) to pass through the given three points. As of this moment the constants a, b, and c have not been assigned those special values. The **solve** command does not result in any values being assigned. Consequently, there are two ways to proceed from here.

We will demonstrate both a conceptually simple (but tedious) way and a subtle way which is operationally very powerful.

First, the tedious way. Just invoke the Maple command **assign** which converts all the = 's in the set q into := 's so that each constant is actually assigned the required value.

```
• assign(q);
```

Note the null return from the **assign** command.

To see the effect of the command, ping Maple for the values of a, b, and c. While you're at it, ping for y too!

```
• a;
```

$$\frac{-16}{15}$$

```
• b;
```

$$\frac{-22}{15}$$

```
• c;
```

$$\frac{51}{5}$$

```
• y;
```

$$-\frac{16}{15}x^2 - \frac{22}{15}x + \frac{51}{5}$$

- `plot(y, x = -8..8);`

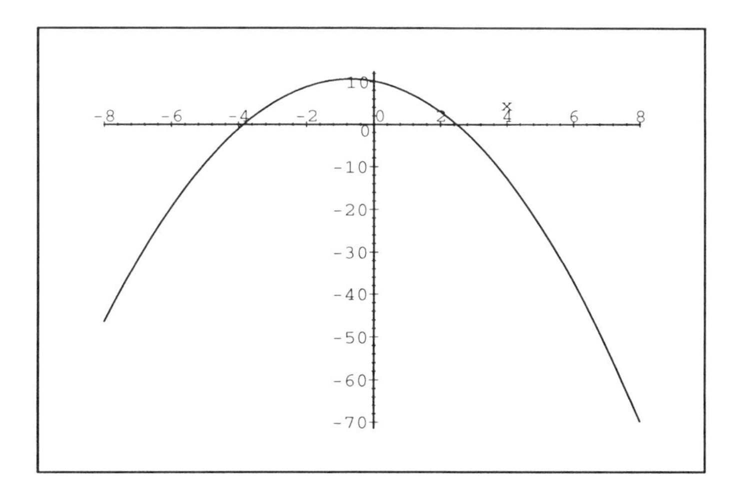

We have accomplished the task. If this were the only interpolation with this parabola, we would stop here. But suppose we wanted to interpolate a different set of three points. To repeat the calculations above, we would first need to "unassign" the values bound to a, b, and c.

Do this by assigning each constant to its "literal" value.

- `a := 'a';`

$$a := a$$

- `b := 'b';`

$$b := b$$

- `c := 'c';`

$$c := c$$

This is what I mean by tedious. Any time variables are assigned values, you run the risk of needing them in their unassigned mode later on. That is why I will show a second style of working that keeps to a minimum the number of variables that get assigned values.

Just for the record, ping y. Notice that once a, b, and c are unassigned, y reverts back to its original general form.

- `y;`

$$a\,x^2 + b\,x + c$$

Now for the subtle approach. Recall that the Maple **subs** command creates a copy of the object into which the substitution is being made. The target expression itself does not change. Hence:

- `yy := subs(q, y);`

$$yy := -\frac{16}{15}x^2 - \frac{22}{15}x + \frac{51}{5}$$

We made three substitutions into the expression for y. These substitutions are the three rules sitting in the set q. By giving the name yy to the result of the substitution command, we have the desired interpolating parabola tagged as yy and the original parabola (1) is still in its general form.

Ping y and yy to verify these claims:

- `y;`

$$a\,x^2 + b\,x + c$$

- `yy;`

$$-\frac{16}{15}x^2 - \frac{22}{15}x + \frac{51}{5}$$

Unit 5: Conic Through Five Points

Five points determine a conic. If you pick five points in the plane, what is the resulting conic that interpolates those points?

The general form of the quadratic expression that determines a conic is

$$ax^2 + bxy + cy^2 + dx + ey + f = 0 \qquad (2)$$

There are 6 constants, not 5. However, at least one has to be nonzero and if (2) is divided through by that nonzero constant, there will be exactly 5 unknowns to be determined. Thus, it generally takes 5 points to determine a conic.

Let's fit a conic through the five points (1,1), (-1,2), (-3, -1), (4,5), and (2,-3).

Enter the lefthand side of equation (2) and give it the name q.

- `q := a*x^2 + b*x*y + c*y^2 + d*x + e*y + f;`

$$q := a\,x^2 + b\,x\,y + c\,y^2 + d\,x + e\,y + f$$

Create two lists of coordinates, X and Y.

- `X := [1, -1, -3, 4,  2];`

$$X := [1, -1, -3, 4, 2]$$

- `Y := [1,  2, -1, 5, -3];`

$$Y := [1, 2, -1, 5, -3]$$

Set up and solve a system of 5 equations in 6 unknowns.

- `for k from 1 to 5 do e.k := subs(x = X[k], y = Y[k], q) = 0; od;`

$$e1 := a + b + c + d + e + f = 0$$

$$e2 := a - 2\,b + 4\,c - d + 2\,e + f = 0$$

$$e3 := 9\,a + 3\,b + c - 3\,d - e + f = 0$$

$$e4 := 16\,a + 20\,b + 25\,c + 4\,d + 5\,e + f = 0$$

$$e5 := 4\,a - 6\,b + 9\,c + 2\,d - 3\,e + f = 0$$

- `q1 := solve({e.(1..5)}, {a,b,c,d,e});`

$$q1 := \left\{b = \frac{929}{3047}\,f, d = -\frac{1393}{3047}\,f, c = \frac{602}{3047}\,f, e = -\frac{1805}{3047}\,f, a = -\frac{1380}{3047}\,f\right\}$$

There are two things to explain here.

First, to avoid typing the string $\{e1, e2, e3, e4, e5\}$, use the shorthand $\{e.(1..5)\}$, which has exactly that same effect.

Second, I assumed above that f was not zero. This is the case if the interpolating conic does not go through the origin. If I had guessed incorrectly, Maple would not have been able to solve the set of equations. Then I would have gone back and excluded some other constant instead of f.

We could have let Maple decide which constant to take as the arbitrary one:

```
solve({e.(1..5)}, {a,b,c,d,e,f});
```

$$\left\{b = \frac{929}{602}\,c,\, d = -\frac{199}{86}\,c,\, f = \frac{3047}{602}\,c,\, e = -\frac{1805}{602}\,c,\, a = -\frac{690}{301}\,c,\, c = c\right\}$$

To extract the information contained in the solution set $q1$, we use the substitution strategy.

```
q2 := subs(q1, q);
```

$$q2 := -\frac{1380}{3047}\,f\,x^2 + \frac{929}{3047}\,f\,x\,y + \frac{602}{3047}\,f\,y^2 - \frac{1393}{3047}\,f\,x - \frac{1805}{3047}\,f\,y + f$$

Setting $f = 3047$ will remove fractions and the last of the unknowns in the equation determining the interpolating conic.

```
q3 := subs(f=3047,q2);
```

$$q3 := -1380\,x^2 + 929\,x\,y + 602\,y^2 - 1393\,x - 1805\,y + 3047$$

We now have two options for obtaining a plot of this conic. We can use Maple's built-in **implicitplot** command or we can solve the quadratic for the individual branches and plot the branches together.

To use the **implicitplot** command, first load the plots package.

```
with(plots):
implicitplot(q3, x = -6..6, y = -10..15);
```

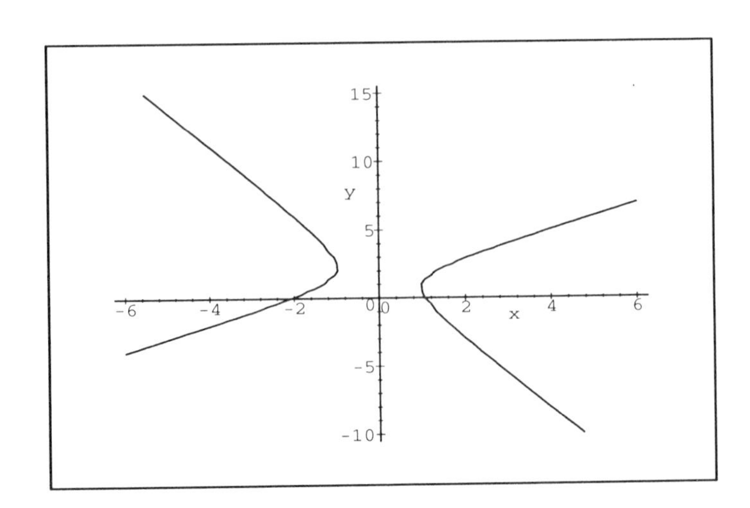

Alternatively, let's solve for the two branches of this conic.

- ```
  q4 := solve(q3 = 0, y);
  ```

$$q4 := -\frac{929}{1204}x + \frac{1805}{1204} + \frac{1}{1204}\sqrt{4186081\,x^2 + 654\,x - 4079151},$$
$$-\frac{929}{1204}x + \frac{1805}{1204} - \frac{1}{1204}\sqrt{4186081\,x^2 + 654\,x - 4079151}$$

The tag $q4$ points to an expression sequence containing the two branches $y1$ and $y2$. for the interpolating conic. Let's just plot these two branches on one set of axes.

- ```
  plot({q4}, x = -6..6);
  ```

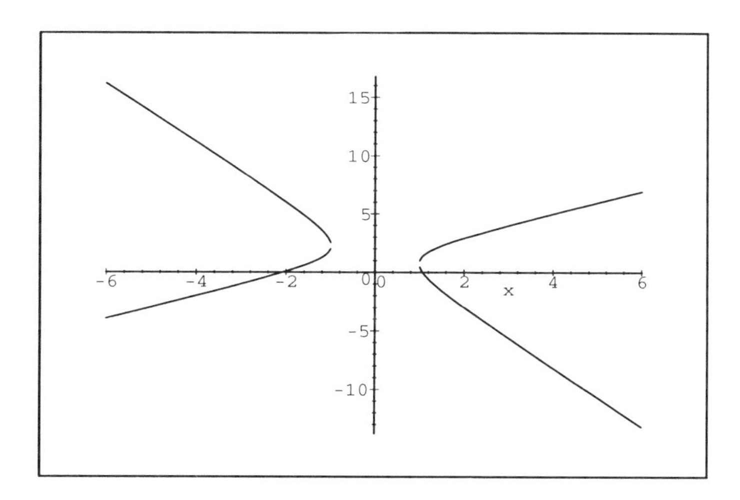

It's the same graph!

Additional activities with conics can be found in the journal article *Exploring Conic Sections with a Computer Algebra System* , R.J. Lopez and J. Mathews, Collegiate Microcomputer, **VIII** (3), August 1990, p. 215-19.

Unit 6: An Implicit Function

Let's see how we might use technology to explore the concept of the implicit function. Given the following quadratic expression q and the equation $q = 0$, an initial attempt at investigating its meaning and behavior might be a call to Maple's **plot** command.

- ```
  q := 17*y^2 + 12*x*y + 8*x^2 - 46*y - 28*x + 17;
  ```

$$q := 17\,y^2 + 12\,x\,y + 8\,x^2 - 46\,y - 28\,x + 17$$

Setting $q = 0$ defines the function $y(x)$ implicitly, but the attempt to plot the expression q will not be successful:

- ```
  plot(q, x = -3..3);
  Warning in iris-plot: empty plot
  ```

Clearly, only functions $y = y(x)$ can be plotted in this fashion. To obtain this form we must solve the equation $q = 0$ for y so that y appears on the left and only x's appear on the right.

- ```
  r := solve(q = 0, y);
  ```

$$r := -\frac{6}{17}\,x + \frac{23}{17} + \frac{2}{17}\sqrt{-25\,x^2 + 50\,x + 60},\\ -\frac{6}{17}\,x + \frac{23}{17} - \frac{2}{17}\sqrt{-25\,x^2 + 50\,x + 60}$$

There are two branches to this implicitly defined function. The two branches are delivered by Maple as an expression sequence. It will be much easier to deal with these two branches if we give them individual names.

- ```
  y1 := r[1];
  ```

$$y1 := -\frac{6}{17}\,x + \frac{23}{17} + \frac{2}{17}\sqrt{-25\,x^2 + 50\,x + 60}$$

- ```
  y2 := r[2];
  ```

$$y2 := -\frac{6}{17}\,x + \frac{23}{17} - \frac{2}{17}\sqrt{-25\,x^2 + 50\,x + 60}$$

Now it should be possible to plot each branch of the implicitly defined function $y(x)$.
  ```

- `plot({y1, y2}, x = -1..3);`

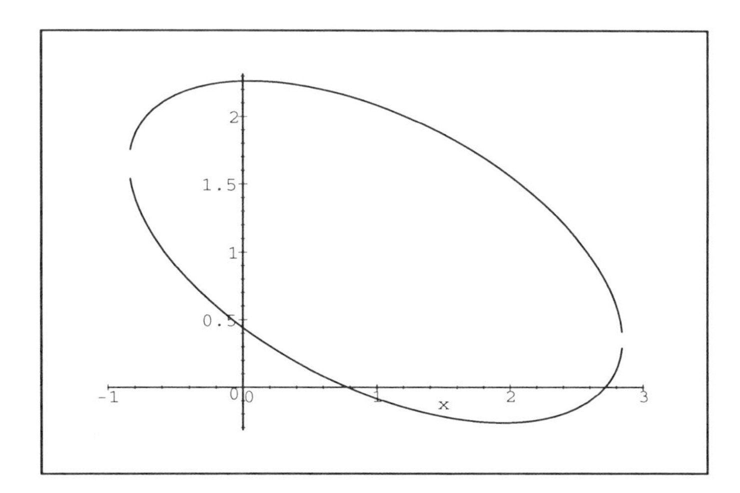

The branches don't quite touch because of the way Maple's **plot** command generates points at which the functions $y1$ and $y2$ are evaluated.

At any rate, the equation $q = 0$ implicitly defines the rotated ellipse just seen.

Of course, we can instead use the **implicitplot** command from the plots package.

- ```
  with(plots):
  implicitplot(q, x = -3..3, y = -1..3);
  ```

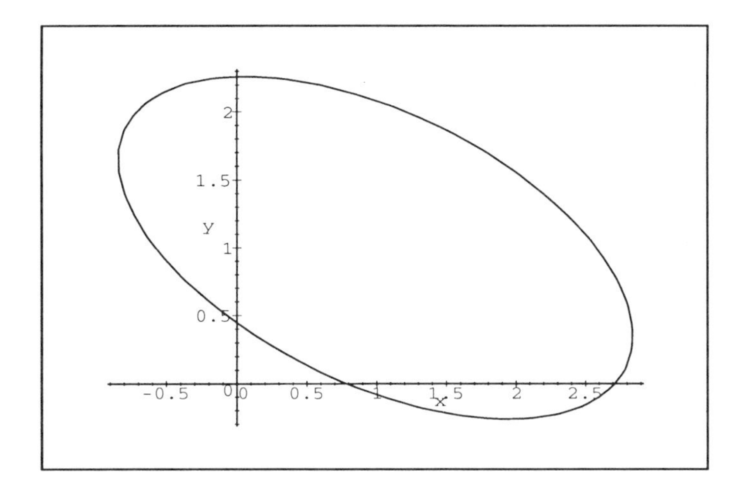

If the lesson is the connection between the implicit representation, the branches, and the algebraic expressions of the branches then the **implicitplot** command does not bring this out by itself. In a learning environment both approaches are probably appropriate for insight.

What else can we do to enrich the mathematical experience just encountered? Here is something I have tried. Introduce a straight line that intersects the ellipse. Calculate the points of intersection of the line and the ellipse.
  ```

- `y3 := x + 1;`

$$y3 := x + 1$$

Now plot the line and the ellipse together on the same set of axes.

- `plot({y1, y2, y3}, x = -1..3);`

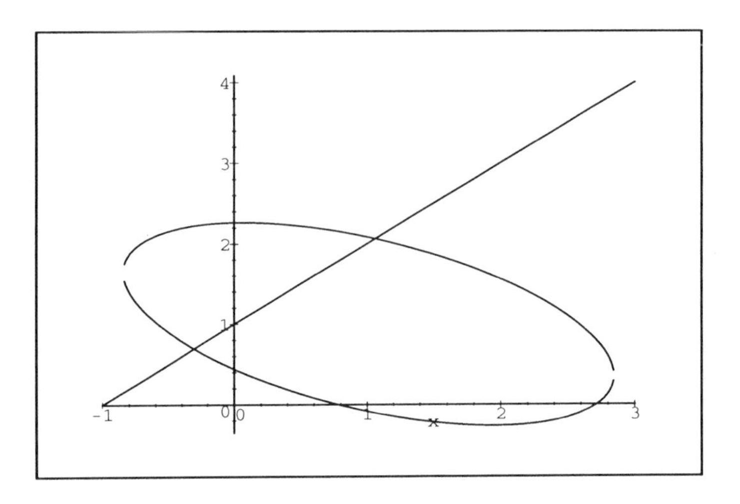

Now do the algebra to solve for the points of intersection. Proceed traditionally: substitute one expression into the other.

- `q1 := subs(y = y3, q);`

$$q1 := 17\,(x+1)^2 + 12\,x\,(x+1) + 8\,x^2 - 74\,x - 29$$

Again, **expand** and **simplify** come to mind. I tried **simplify** and it worked.

- `q2 := simplify(q1);`

$$q2 := 37\,x^2 - 28\,x - 12$$

Solving the equation $q2 = 0$ for x would lead to the discovery of the coordinates of the two points of intersection. This involves activities like those found in Unit 1. Rather than repeat them, let's pose a different question.

What is the domain of for each branch of the implicitly defined function $y(x)$? This question can be answered if we deduce which values for x keep the square root in either branch real.

- `y1;`

$$-\frac{6}{17}\,x + \frac{23}{17} + \frac{2}{17}\,\sqrt{-25\,x^2 + 50\,x + 60}$$

How many "pieces," or terms, does Maple see in the expression for $y1$?

- `nops(y1);`

$$3$$

Maple is claiming that $y1$ has three pieces. What are they?

- `op(y1);`

$$-\frac{6}{17}\,x,\frac{23}{17},\frac{2}{17}\,\sqrt{-25\,x^2+50\,x+60}$$

The individual pieces of $y1$ are listed in an expression sequence. The square root, the one we want, is the third one. Thus,

- `q3 := op(3, y1);`

$$q3 := \frac{2}{17}\,\sqrt{-25\,x^2+50\,x+60}$$

We have assigned to the tag $q3$ a multiple of the square root we need. We can play the **nops/op** game further if we need to, but here we are going to solve the equation $q3 = 0$ for x. The constant multipliers in front of the square root will not matter.

- `q4 := solve(q3 = 0, x);`

$$q4 := 1-\frac{1}{5}\,\sqrt{85},1+\frac{1}{5}\,\sqrt{85}$$

As floating point numbers, the endpoints of the interval are

- `evalf(q4);`

$$-.843908891,2.843908891$$

We have found the desired delimiters of the domain of the branches of the implicitly defined function $y(x)$.

Unit 7: Inverse Functions

Here is a lesson designed to illuminate the notion of an inverse function. Let's graph, and then compute, the inverse of a function $f(x)$.

To obtain the inverse function, first solve the equation $f(x) = y$ for the variable x , thereby producing the function $x = g(y)$. Then switch the letters to obtain $y = g(x)$.

Graphically, this amounts to reflecting the graph of $f(x)$ about the line $y = x$. A plot showing these relationships is a must!

- ```
  f := 1 + sqrt(1 + x^2);
  ```

$$f := 1 + \sqrt{1 + x^2}$$

- ```
  plot(f, x = -1..1);
  ```

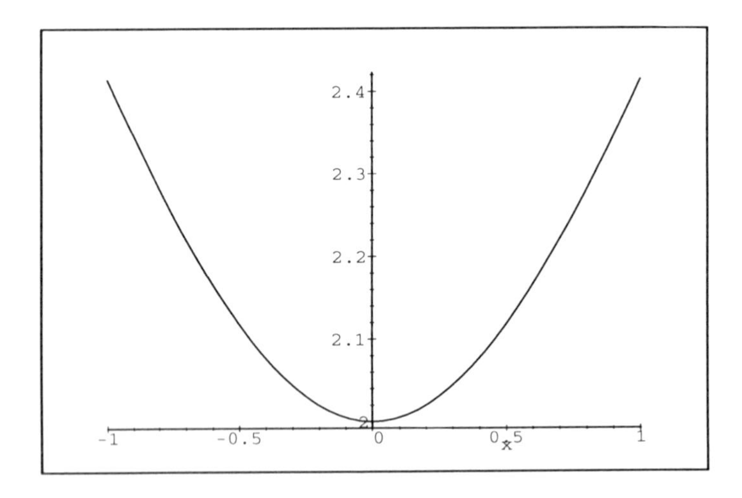

From the graph we see that $f(x)$ is not a monotone function. To "invert" $f(x)$, we will have to select a monotone portion and work with that new function. For simplicity, let's choose the branch for which x is nonnegative.

If $f(x)$ is the function with ordered pairs $(x, f(x))$, then the inverse function has ordered pairs $(f(x), x)$. We can exploit this relationship and parametrically plot both $f(x)$ and its inverse $g(x)$ before doing any of the algebra needed for actually finding $g(x)$.

- `plot({[x, f, x = 0..3], [f, x, x = 0..4]});`

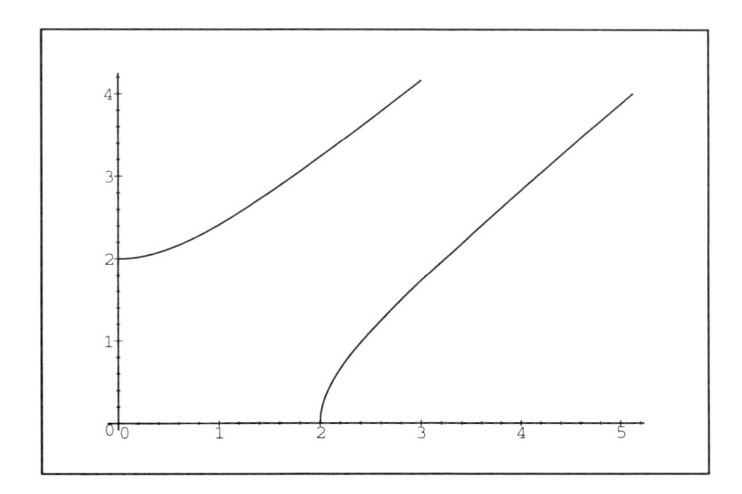

Why not include the line $y = x$?

- `plot({[x, f, x = 0..3], [f, x, x = 0..4], [x, x, x = 0..4]});`

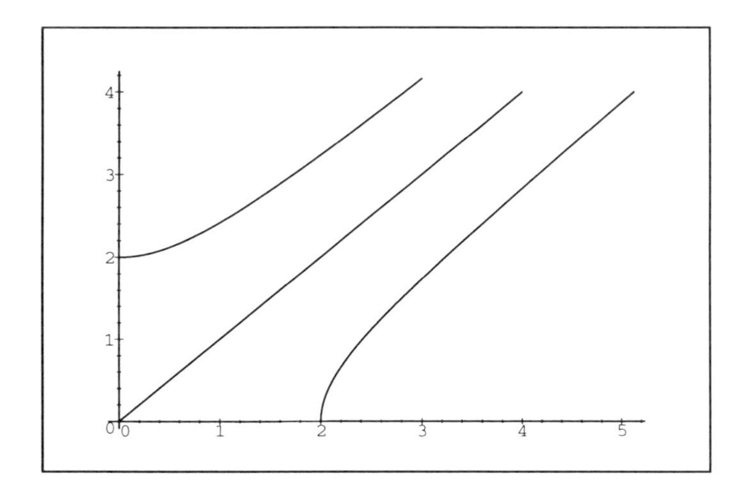

This is the graph we wanted, and we got it without creating the expression for $g(x)$, thereby illustrating the ordered-pair interpretation of a function. Now, we can do the analysis.

- `q := solve(f = y, x);`

$$q := \sqrt{-1+(-1+y)^2}, -\sqrt{-1+(-1+y)^2}$$

As expected, there are two solutions, only one of which corresponds to the nonnegative branch. Hence, we want the first solution. The inverse $g(x)$ is the first of the members of the expression sequence q, provided we replace the y with x.

- `g := subs(y = x, q[1]);`

$$g := \sqrt{-1 + (-1 + x)^2}$$

We can again create the plot of $f(x)$, $g(x)$, and the line $y = x$.

- `plot({[x, f, x = 0..3], [x, g, x = 2..7/2], [x, x, x = 0..4]});`

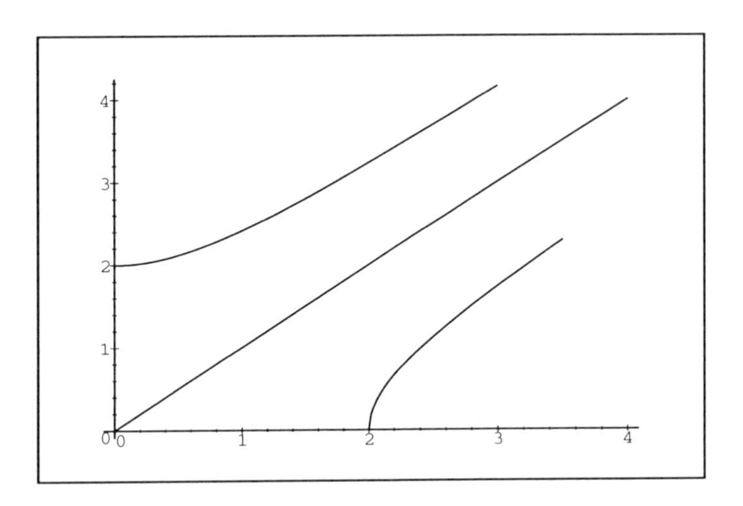

We conclude that since this graph and the earlier graph in which the ordered pairs were reversed are the same, the effect of reversing the ordered pairs is to create the inverse function. Moreover, analytically, we can verify that $f(g(x)) = g(f(x)) = x$, the defining relationship for inverse functions. In Release 3, Maple no longer simplifies

- `sqrt(x^2);`

$$\sqrt{x^2}$$

to x unless Maple knows enough about x to make this a true statement. Given the nature of $f(x)$ and $g(x)$, we anticipate Maple's need to know that x is positive.

- `assume(x>0);`

Then, we compute $g(f(x)) = x$ via

- `simplify(subs(x = f, g));`

$$x\tilde{}$$

The tilde (˜) attached to the x is Maple's way of reminding us that the result is true only in light of our assumption made about x.

To compute $f(g(x)) = x$ we try

- `simplify(subs(x = g, f));`

$$1 - \text{signum}(-1 + x\tilde{}) + \text{signum}(-1 + x\tilde{})\, x\tilde{}$$

Ah, yes. The domain of $g(x)$ is not the same as the domain of $f(x)$. In fact, from the graphs above we see we need to

- `assume(x>1);`

Then, we obtain $f(g(x)) = x$ via

- `simplify(subs(x = g, f));`

$$x^{\sim}$$

An assumption on x can be removed in the same way that one would release x from any other assigned value:

- `x := 'x';`

$$x := x$$

Unit 8: Partial Fraction Decomposition

The partial fraction decomposition typically arises as a technique of integration. Some students see the technique again when inverting Laplace transforms. Nevertheless, the manipulations are algebraic, not analytic. Hence, we present partial fraction decomposition as an algebraic operation, divorced from its use in the applications.

The following nasty fraction was actually given on a final exam, but with Maple the students were successful in replicating the steps of a "by-hand" implementation of partial fractions. This unit illustrates Maple reproducing such a calculation.

- ```
  f := (2*x^7 + 11*x^6 - 12*x^5 - 93*x^4 - 70*x^3 + 91*x^2 +
                      43*x - 34) /
       (x^6 + 4*x^5 - 12*x^4 - 34*x^3 + 12*x^2 + 32*x - 15);
  ```

$$f := \frac{2\,x^7 + 11\,x^6 - 12\,x^5 - 93\,x^4 - 70\,x^3 + 91\,x^2 + 43\,x - 34}{x^6 + 4\,x^5 - 12\,x^4 - 34\,x^3 + 12\,x^2 + 32\,x - 15}$$

This would be brutal to do "by hand." Although Maple will deliver the decomposition with a single command, as below, this unit illustrates the implementation of a standard "by-hand" algorithm in Maple.

- ```
  convert(f, parfrac, x);
  ```

$$2\,x + 3 - 2\,\frac{1}{x+5} + \frac{1}{x-3} + \frac{-1+x}{x^2+x-1} + \frac{-1+2\,x}{(\,x^2+x-1\,)^2}$$

This is the decomposition. But unless we dig a bit deeper we might miss some of the salient algebraic milestones present. For example, the fraction is "improper," so we need to perform a long division.

- ```
  quo(numer(f), denom(f), x, 'r');
  ```

$$2\,x + 3$$

The **quo** command delivers the quotient and stores the remainder as the variable r. The quotes on the letter r are important if r has a previously defined value.

The actual fraction to be decomposed is the remainder r divided by the denominator of f.

- ```
  q := r/denom(f);
  ```

$$q := \frac{11\,x^4 + 8\,x^3 - 9\,x^2 - 23\,x + 11}{x^6 + 4\,x^5 - 12\,x^4 - 34\,x^3 + 12\,x^2 + 32\,x - 15}$$

The factors of the denominator can be read from the original answer that Maple produced, or else they can be found with the **factor** command.

- `factor(denom(f));`

$$(x+5)(x-3)(x^2+x-1)^2$$

There are four individual fractions needed for this decomposition. They are as follows.

- `f1 := (a*x + b) / (x^2 + x - 1)^2;`

$$f1 := \frac{a\,x+b}{(x^2+x-1)^2}$$

- `f2 := (c*x + d) / (x^2 + x - 1);`

$$f2 := \frac{c\,x+d}{x^2+x-1}$$

- `f3 := e / (x + 5);`

$$f3 := \frac{e}{x+5}$$

- `f4 := g / (x - 3);`

$$f4 := \frac{g}{x-3}$$

The algorithm calls for the addition of these four fractions.

- `s := f1 + f2 + f3 + f4;`

$$s := \frac{a\,x+b}{(x^2+x-1)^2} + \frac{c\,x+d}{x^2+x-1} + \frac{e}{x+5} + \frac{g}{x-3}$$

Incidentally, this addition could have been accomplished by using Maple's **sum** command. One thing to watch out for is Maple's unhappiness with concatenated names in the **sum** command. The single (forward) quotes used below cure this unhappiness.

- `sum('f.k', 'k' = 1..4);`

$$\frac{a\,x+b}{(x^2+x-1)^2} + \frac{c\,x+d}{x^2+x-1} + \frac{e}{x+5} + \frac{g}{x-3}$$

At any rate, we need Maple to add these fractions!

- `s1 := normal(s);`

$$\begin{aligned} s1 := (&15\,d - 3\,e - 15\,b + 5\,g - 15\,a\,x + 15\,c\,x + 2\,a\,x^2 + 2\,b\,x + a\,x^3 \\ &+ b\,x^2 - 17\,c\,x^2 - 17\,d\,x - 14\,c\,x^3 - 14\,d\,x^2 + c\,x^5 + 3\,c\,x^4 + d\,x^4 \\ &+ 3\,d\,x^3 + e\,x^5 - e\,x^4 - 7\,e\,x^3 + e\,x^2 + 7\,e\,x + g\,x^5 + 7\,g\,x^4 + 9\,g\,x^3 \\ &- 7\,g\,x^2 - 9\,g\,x) \Big/ \left((x+5)(x-3)(x^2+x-1)^2\right) \end{aligned}$$

The numerators for both forms of the fraction q must match identically. This means that $q = s1$ must be an identity in x, and like powers in the numerators must have equal coefficients. We'll use this strategy to form equations for $a, b, c, d, e,$ and g.

- ```
  for k from 0 to 5 do
  e.k := coeff(numer(q), x, k) = coeff(numer(s1), x, k);
  od;
  ```

$$e0 := 11 = 15\,d - 3\,e - 15\,b + 5\,g$$
$$e1 := -23 = -15\,a + 15\,c - 9\,g + 2\,b + 7\,e - 17\,d$$
$$e2 := -9 = e + 2\,a - 14\,d - 17\,c + b - 7\,g$$
$$e3 := 8 = -14\,c - 7\,e + 3\,d + a + 9\,g$$
$$e4 := 11 = 3\,c - e + d + 7\,g$$
$$e5 := 0 = c + e + g$$

We now seek a simultaneous solution of these six equations in six unknowns.

- ```
  q1 := solve({e.(0..5)}, {a,b,c,d,e,g});
  ```

$$q1 := \{ e = -2, g = 1, b = -1, a = 2, d = -1, c = 1 \}$$

The set $q1$ contains the replacements that should be made in the fractions $f1$, $f2$, $f3$, and $f4$. An efficient way to apply these replacements without formally assigning values to variables is to use substitution.

- ```
  for k from 1 to 4 do F.k := subs(q1, f.k); od;
  ```

$$F1 := \frac{-1 + 2\,x}{(\,x^2 + x - 1\,)^2}$$
$$F2 := \frac{-1 + x}{x^2 + x - 1}$$
$$F3 := -2\,\frac{1}{x + 5}$$
$$F4 := \frac{1}{x - 3}$$

The astute reader will notice, perhaps, that the quadratic factor in the example above is actually factorable over the reals. Indeed,

- ```
  solve(x^2 + x - 1, x);
  ```

$$-\frac{1}{2} + \frac{1}{2}\sqrt{5}, -\frac{1}{2} - \frac{1}{2}\sqrt{5}$$

This raises a question among Electrical Engineers, for example, who prefer that partial fraction decompositions be performed over the complex field so that all factors are linear! Can Maple be induced to factor a polynomial into linear factors over the complex number field? The answer, of course, is "yes," but it does take some extra manipulations to implement this extra requirement.

First, the zeros of the polynomial in the denominator must be calculated.

- ```
  R := solve(denom(f), x);
  ```

$$R := -5, 3, -\frac{1}{2} + \frac{1}{2}\sqrt{5}, -\frac{1}{2} - \frac{1}{2}\sqrt{5}, -\frac{1}{2} + \frac{1}{2}\sqrt{5}, -\frac{1}{2} - \frac{1}{2}\sqrt{5}$$
  ```

Next, the linear factors need to be constructed from these roots.

- `F := product((x - R[j]), j = 1..6);`

$$F := (x+5)(x-3)\left(x+\frac{1}{2}-\frac{1}{2}\sqrt{5}\right)^2\left(x+\frac{1}{2}+\frac{1}{2}\sqrt{5}\right)^2$$

Since we are only alerting the reader to the possibilities contained here, we'll simply show that the desired decomposition now follows.

- `convert(numer(f)/F, parfrac, x);`

$$2x+3-2\frac{1}{x+5}+\frac{1}{x-3}$$
$$+\frac{1}{50}\frac{\left(37088\sqrt{5}-58976\right)\sqrt{5}}{\left(54-14\sqrt{5}\right)\left(86+18\sqrt{5}\right)\left(2x+1-\sqrt{5}\right)}$$
$$+\frac{1}{20}\frac{-992\sqrt{5}+2016}{\left(\sqrt{5}-7\right)\left(9+\sqrt{5}\right)\left(2x+1-\sqrt{5}\right)^2}$$
$$+\frac{1}{50}\frac{\left(37088\sqrt{5}+58976\right)\sqrt{5}}{\left(54+14\sqrt{5}\right)\left(86-18\sqrt{5}\right)\left(2x+1+\sqrt{5}\right)}$$
$$-\frac{1}{20}\frac{-992\sqrt{5}-2016}{\left(\sqrt{5}-9\right)\left(\sqrt{5}+7\right)\left(2x+1+\sqrt{5}\right)^2}$$

Unit 9: Derivatives By Definition

An essential step in acquiring mastery of calculus is seeing the derivative as the limit of the slopes of secant lines. This means seeing the derivative as the limit of a difference quotient. One of the obstacles to this understanding is the algebra needed to take the various limits that arise in this context. Let's explore how Maple can be used to learn about the definition of the derivative without requiring the student to be an algebraic virtuoso. We'll pick a function, set up the appropriate difference quotient, and explore the idea of the limit.

- `f := 3*x^2 - 5*x + 2;`

$$f := 3\,x^2 - 5\,x + 2$$

First, let's create a difference quotient and derivative at the specific point $(2, f(2))$.

- `m := (subs(x = 2 + h, f) - subs(x = 2, f))/h;`

$$m := \frac{3\,(2+h)^2 - 12 - 5\,h}{h}$$

The derivative at $x = 2$ is the value of the slope m when "h is zero." What happens if we simply substitute $h = 0$ in m?

- `subs(h = 0, m);`

Error, division by zero

We are at the crux of the limit issue. We can't let $h = 0$ in m, but somehow, as h gets small, m has a meaning. When in doubt, draw a graph.

- `plot(m, h = 0..1, 'm' = 0..10);`

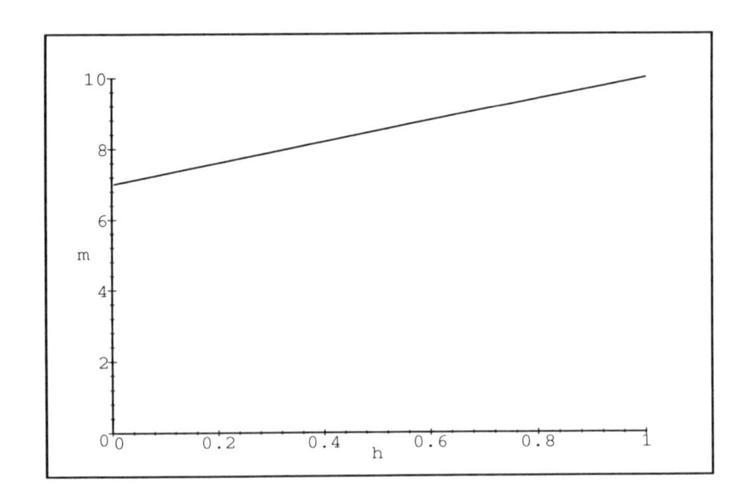

For small values of h, the slope m approaches 7. But it is not appropriate to substitute 0 for h in the expression for m . What happens if this expression is modified by simplifying it?

- `m1 := simplify(m);`

$$m1 := 7 + 3\,h$$

Ah! The simplified version of m is nothing more than the equation of a line. From the expression $m1$ we can see that for $h = 0$ the value of $m1$ is just 7. So, except for the one value $h = 0$, the expression for m is really the expression for the straight line represented by $m1$. Is there a mathematical construct to deal with this?

- `limit(m, h = 0);`

$$7$$

The Maple command **limit** does the appropriate examination of the expression for the slope and finds out what happens to it as h gets small.

We next do these calculations at the arbitrary point $(a, f(a))$.

- `m := (subs(x = a + h, f) - subs(x = a, f))/h;`

$$m := \frac{3\,(a+h)^2 - 5\,h - 3\,a^2}{h}$$

- `limit(m, h = 0);`

$$6\,a - 5$$

At any point $(a, f(a))$ the slopes of the secant lines have limiting value $6a - 5$. Since a was a place-holder for a value of the x-coordinate, we should be able to reproduce this calculation in terms of x itself.

- `m := (subs(x = x + h, f) - f)/h;`

$$m := \frac{3\,(x+h)^2 - 5\,h - 3\,x^2}{h}$$

- `limit(m, h = 0);`

$$6\,x - 5$$

The expression $6x - 5$ is the limit of the difference quotient at x. The mathematical name for this number is *derivative*. You might guess that Maple has a built-in command for generating this limit of the difference quotient.

- `diff(f, x);`

$$6\,x - 5$$

The **diff** command calculates the "derivative" of f(x). The process of computing a derivative is called *differentiation*, and that is where the notation **diff** comes from.

The derivative is the limiting slope of secant lines on the graph of $f(x)$. Clearly the secant line becomes a tangent line as h approaches 0. Thus, the derivative is the slope of the tangent line.

Let's construct, at the point $(2,4)$, the line tangent to the graph of $f(x)$.

- `d := diff(f, x);`

$$d := 6\,x - 5$$

The slope of this tangentline is the value of the derivative at $x = 2$.

- `m := subs(x = 2, d);`

$$m := 7$$

The equation of the tangent line is therefore

- `y := m*(x - 2) + 4;`

$$y := 7\,x - 10$$

A graph of this tangent line and of the function f is created via

- `plot({f, y}, x = 0..3);`

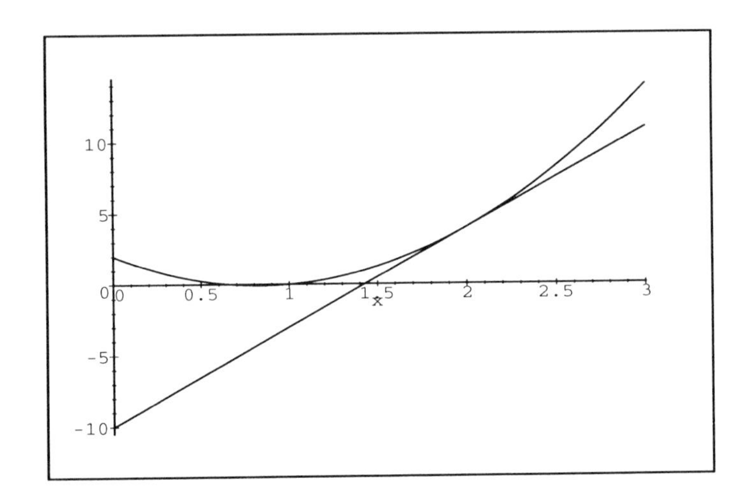

A nicer plot can be obtained by representing the curves parametrically:

```
• plot({[x, f, x = -2..4], [x, y, x = 1..3]});
```

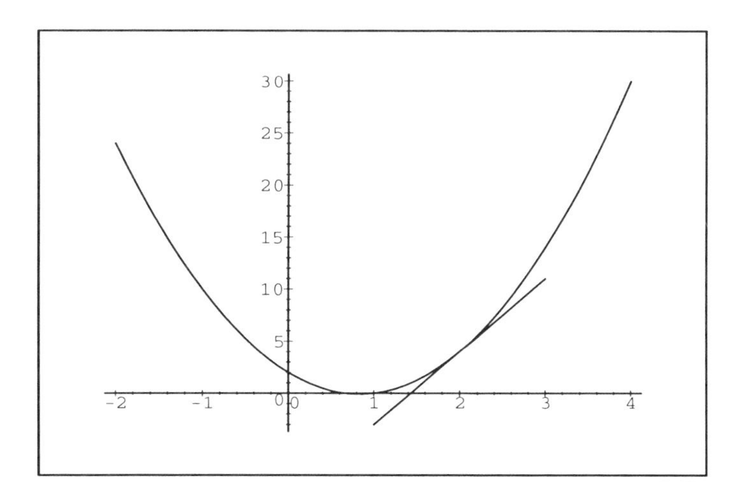

In either event, we see that the line through $(2, 4)$ with slope 7, as calculated from the derivative, is tangent to the graph of $f(x)$.

Finally, let's explore the calculation of a derivative for some other function.

```
• f := sin(x);
```

$$f := \sin(x)$$

The difference quotient (which represents the slopes of secant lines) is

```
• m := (subs(x = x + h, f) - f)/h;
```

$$m := \frac{\sin(x + h) - \sin(x)}{h}$$

The limit of the difference quotient is found by the computation

```
• limit(m, h = 0);
```

$$\cos(x)$$

Finally, we compare this result with a direct call to the differentiation command.

```
• diff(f, x);
```

$$\cos(x)$$

It would have been a surprise if $\cos(x)$ not been found by both methods!

Here is another exploration sometimes used when teaching the derivative. The difference quotient at $(x, f(x))$ is treated as a function of x. This means h is a parameter and the difference quotient represents a family of curves that should converge to the derivative as h goes to 0.

Let's examine this idea using the function $f(x) = \ln(x)$, where "ln" is the natural logarithm.

- `f := ln(x);`

$$f := \ln(x)$$

The difference quotient is given as

- `m := (subs(x = x + h, f) - f)/h;`

$$m := \frac{\ln(x+h) - \ln(x)}{h}$$

Create some specific members of the family of curves represented by the difference quotient m. Do this by selecting specific values of h.

- `for k from 0 to 5 do m.k := subs(h = 1/2^k, m); od;`

$$m0 := \ln(x+1) - \ln(x)$$

$$m1 := 2\ln\left(x + \frac{1}{2}\right) - 2\ln(x)$$

$$m2 := 4\ln\left(x + \frac{1}{4}\right) - 4\ln(x)$$

$$m3 := 8\ln\left(x + \frac{1}{8}\right) - 8\ln(x)$$

$$m4 := 16\ln\left(x + \frac{1}{16}\right) - 16\ln(x)$$

$$m5 := 32\ln\left(x + \frac{1}{32}\right) - 32\ln(x)$$

A plot of these functions, which represent difference quotients for various values of h, shows convergence to the limiting function that is the derivative.

- `plot({m.(0..5)}, x = 0..3, 0..3);`

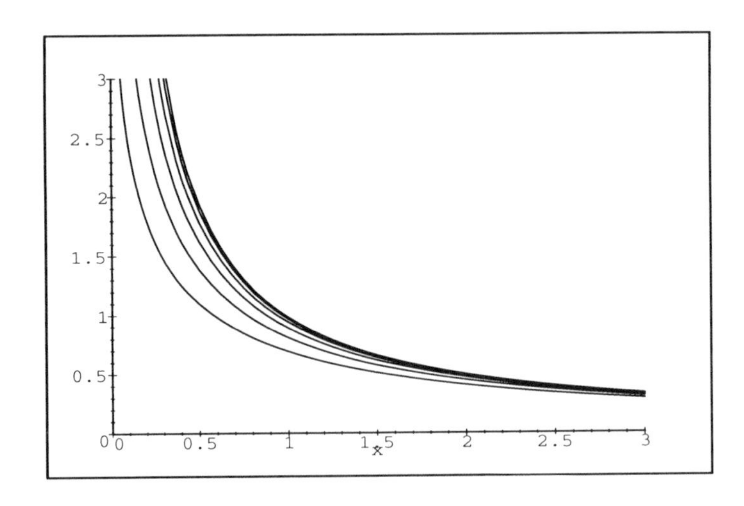

Unit 10: Implicit Differentiation

In this unit we implement implicit differentiation in a manner consistent with a conceptual approach to the topic. Essentially, we stress that each appearance of the dependent variable y is to be interpreted as $y(x)$, the implicitly defined function whose derivative we seek.

Let $f(x, y) = a$ define $y = y(x)$ implicitly. For example, take

- `f := x*sin(y) + (x*y^2 - 1)/y = a;`

$$f := x\sin(y) + \frac{x\,y^2 - 1}{y} = a$$

To emphasize that y really represents the implicitly defined $y(x)$, replace each y in $f(x, y)$ with the more explicit $y(x)$.

- `q := subs(y = y(x), f);`

$$q := x\sin(\mathrm{y}(x)) + \frac{x\,\mathrm{y}(x)^2 - 1}{\mathrm{y}(x)} = a$$

Now apply the differentiation operation to both sides of the equation q.

- `q1 := diff(q, x);`

$$q1 := \sin(\mathrm{y}(x)) + x\cos(\mathrm{y}(x))\left(\frac{\partial}{\partial x}\mathrm{y}(x)\right) + \frac{\mathrm{y}(x)^2 + 2\,x\,\mathrm{y}(x)\left(\frac{\partial}{\partial x}\mathrm{y}(x)\right)}{\mathrm{y}(x)} - \frac{(x\,\mathrm{y}(x)^2 - 1)\left(\frac{\partial}{\partial x}\mathrm{y}(x)\right)}{\mathrm{y}(x)^2} = 0$$

Solve this equation for the desired derivative $\frac{dy}{dx}$. Refer to this derivative in Maple by its full name: **diff (y (x), x)**.

- `q2 := solve(q1, diff(y(x),x));`

$$q2 := -\frac{\sin(\mathrm{y}(x)) + \mathrm{y}(x)}{x\cos(\mathrm{y}(x)) + x + \dfrac{1}{\mathrm{y}(x)^2}}$$

That's the implicit derivative. It is both convenient and usual to restore the symbol $y(x)$ to the simpler y.

- `d := subs(y(x) = y, q2);`

$$d := -\frac{\sin(y) + y}{x\cos(y) + x + \dfrac{1}{y^2}}$$

Unit 11: Taylor Polynomials

To learn about Taylor polynomials,we will relegate to Maple the computations needed and attempt to show some activities by which insight into the behavior of these very important objects might be obtained. Let's enter a function, create some of its Taylor polynomials, and then answer at least one non-trivial question about these polynomials.

For the purist, we will use Maple to demonstrate the mechanics of finding the Taylor expansion.

- `f := 100*arctan(5*x^3 - 3*x^2 - 4*x + 6);`

$$f := 100 \arctan(5\,x^3 - 3\,x^2 - 4\,x + 6)$$

- `plot(f, x=-1..1);`

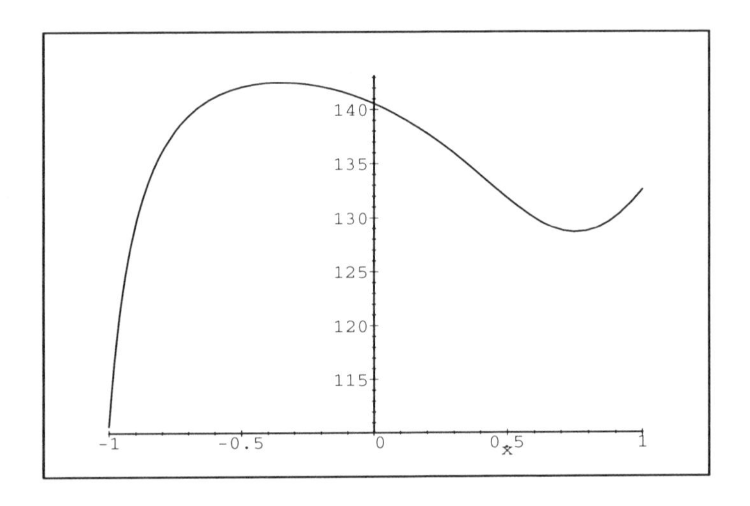

Maple has abbreviated the y-axis in order to show more detail where the function is of more interest.

Now, at $x = 0$, create $p3$, the Taylor polynomial of degree 3.

- `g := taylor(f, x = 0, 4);`

$$g := 100 \arctan(6) - \frac{400}{37}\,x - \frac{20700}{1369}\,x^2 - \frac{229700}{151959}\,x^3 + \mathrm{O}(x^4)$$

The syntax for the **taylor** command allows an integer, here 4, for the order of the *error term*, given by the *BIG OH* symbol. It is imperative that this symbol be removed from the expression that seems to be the Taylor polynomial. The return of the **taylor** command is a "series" data structure. If this data structure is not converted to a polynomial data structure, nothing of any use can be done with the initial return of the **taylor** command.

- `g1 := convert(g, polynom);`

$$g1 := 100 \arctan(6) - \frac{400}{37}\,x - \frac{20700}{1369}\,x^2 - \frac{229700}{151959}\,x^3$$

As promised, this form is a true polynomial data structure that can be plotted, evaluated, etc. We'll plot the function and this third-degree Taylor polynomial.

- `plot({f, g1}, x = -1..1);`

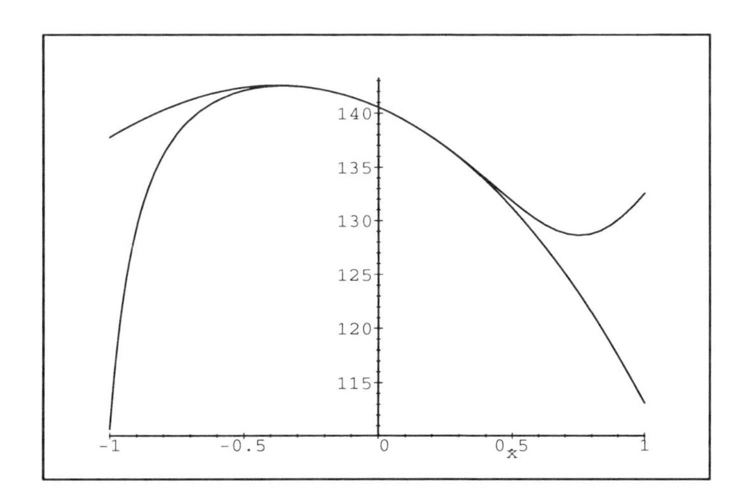

Before we investigate questions such as how well does the Taylor polynomial approximate the function, or over what domain is the polynomial accurate, we illustrate the algorithm by which the Taylor polynomial is constructed. Let's create this same Taylor polynomial "by hand" with Maple performing the necessary steps at our direction. The term of order zero is the value of the function at the point of the expansion.

- `subs(x = 0, f);`

$$100 \arctan(6)$$

We have just obtained the constant term of the polynomial $p3$. The remaining coefficients are built from derivatives. We need the first three derivatives of f. How about using a Maple loop?

- `for k from 1 to 3 do d.k := diff(f, x$k); od;`

$$d1 := 100 \frac{15x^2 - 6x - 4}{1 + (5x^3 - 3x^2 - 4x + 6)^2}$$

$$d2 := 100 \frac{30x - 6}{1 + (5x^3 - 3x^2 - 4x + 6)^2} - 200 \frac{(15x^2 - 6x - 4)^2 (5x^3 - 3x^2 - 4x + 6)}{(1 + (5x^3 - 3x^2 - 4x + 6)^2)^2}$$

$$d3 := 3000 \frac{1}{1 + \%1^2} - 600 \frac{(30x - 6)\,\%1\,(15x^2 - 6x - 4)}{(1 + \%1^2)^2} + 800 \frac{(15x^2 - 6x - 4)^3 \%1^2}{(1 + \%1^2)^3} - 200 \frac{(15x^2 - 6x - 4)^3}{(1 + \%1^2)^2}$$

$$\%1 := 5x^3 - 3x^2 - 4x + 6$$

First, note the use of the notation "x$k" in the call to the **diff** function. This is one form of Maple's sequence operator, and when $k = 3$, it produces the string "x, x, x." In addition, note Maple's use of the "%1" notation. Since there are clusters of symbols that appear more than once, Maple, just as any human would, uses an abbreviation for such clumps.

But we need to evaluate these derivatives at $x = 0$. We also might as well toss in division by the appropriate factorial in each case.

- ```
  for k from 1 to 3 do subs(x = 0, d.k) / k! ; od;
  ```

$$\frac{-400}{37}$$
$$\frac{-20700}{1369}$$
$$\frac{-229700}{151959}$$

If we compare these values to the coefficients of $p3$, we realize that we have, in essence, constructed the Taylor polynomial for ourselves.

- ```
  g1;
  ```

$$100\arctan(6) - \frac{400}{37}x - \frac{20700}{1369}x^2 - \frac{229700}{151959}x^3$$

Clearly, activities such as these clarify the mechanics of creating Taylor polynomials. With Maple, however, we should be led to explore insights into the behavior of the Taylor polynomial.

Let's create the first 9 Taylor polynomials at $x = -.6$ by writing an appropriate Maple loop. In this loop we will **convert** immediately from series, to polynomial, data structures.

- ```
  for k from 0 to 8 do
      p.k := convert(taylor(f, x = -.6, k+1), polynom);
  od;
  ```

$$p0 := 141.1891065$$

$$p1 := 148.7008248 + 12.51953047\,x$$

$$p2 := 148.7008248 + 12.51953047\,x - 39.82736445\,(x + .6)^2$$

$$p3 := 148.7008248 + 12.51953047\,x - 39.82736445\,(x + .6)^2 + 67.04118176\,(x + .6)^3$$

$$p4 := 148.7008248 + 12.51953047\,x - 39.82736445\,(x + .6)^2 + 67.04118176\,(x + .6)^3 - 136.2544836\,(x + .6)^4$$

$$p5 := 148.7008248 + 12.51953047\,x - 39.82736445\,(x + .6)^2 + 67.04118176\,(x + .6)^3 - 136.2544836\,(x + .6)^4 + 260.8288496\,(x + .6)^5$$

$$p6 := 148.7008248 + 12.51953047\,x - 39.82736445\,(x + .6)^2 + 67.04118176\,(x + .6)^3 - 136.2544836\,(x + .6)^4 + 260.8288496\,(x + .6)^5 - 504.2656599\,(x + .6)^6$$
  ```

$$p7 := 148.7008248 + 12.51953047\,x - 39.82736445\,(x + .6)^2 + 67.04118176\,(x + .6)^3 - 136.2544836\,(x + .6)^4 + 260.8288496\,(x + .6)^5 - 504.2656599\,(x + .6)^6 + 970.1546661\,(x + .6)^7$$

$$p8 := 148.7008248 + 12.51953047\,x - 39.82736445\,(x + .6)^2 + 67.04118176\,(x + .6)^3 - 136.2544836\,(x + .6)^4 + 260.8288496\,(x + .6)^5 - 504.2656599\,(x + .6)^6 + 970.1546661\,(x + .6)^7 - 1859.958656\,(x + .6)^8$$

The desire to plot these polynomials along with the function $f(x)$ is universal.

- `plot({f, p.(0..6)}, x = -1..1, 100..150);`

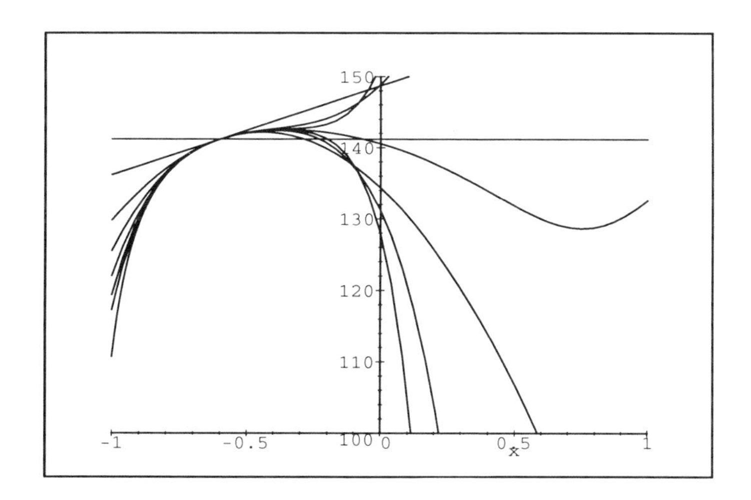

Here are two questions that could be posed.

1. What is the lowest ordered Taylor polynomial for $f(x)$, generated at $x = a$, that approximates $f(b)$ to within some arbitrarily small number called ϵ?

2. Over what interval I, symmetrically centering $x = a$, does the n-th Taylor polynomial of $f(x)$ at $x = a$ approximate $f(x)$ to within some arbitrarily small number called ϵ?

The first question fixes the "interval" and looks for the degree n; the second, fixes the degree n and seeks an interval I.

Consider the first question. For simplicity, use $f(x)$ itself to generate the "exact" value for $f(-.3)$. Then, have Maple compare this value to the value of each polynomial at $x = -.3$.

- `y := evalf(subs(x = -.3, f));`

$$y := 142.4678150$$

Now, loop through all possible cases.

- `for k from 0 to 8 do k, abs(y - subs(x = -.3, p.k)); od;`

$$0, 1.2787085$$
$$1, 2.4771507$$
$$2, 1.1073121$$
$$3, .7027998$$
$$4, .4008615$$
$$5, .2329526$$
$$6, .1346571$$
$$7, .0775157$$
$$8, .0445162$$

If ϵ were $.1$, then from the table above we see that $p7$ is the first polynomial that approximates $f(-.3)$ to within ϵ.

Unit 12: Teaching the Definite Integral

Maple contains features both useful and effective for seeing the definite integral as area under a curve. We begin by posing the question: How can we find the area bounded by the graph of some function $f(x)$ and the x-axis? We are then led to an approximate answer based on rectangles drawn under the graph of $f(x)$.

Maple is "modularized" for efficiency. It contains command clusters called packages, one of which caters to needs of students and is therefore called the **student** package. We load this package to access those of its commands needed in this exploration.

- `with(student);`

$$[D, Doubleint, Int, Limit, Lineint, Sum, Tripleint, changevar, combine, completesquare, distance, equate, extrema, integrand, intercept, intparts, isolate, leftbox, leftsum, makeproc, maximize, middlebox, middlesum, midpoint, minimize, powsubs, rightbox, rightsum, showtangent, simpson, slope, trapezoid, value]$$

By terminating the command **with(student)** with the semicolon, the complete output from Maple is visible. In the displayed list we can see all the commands now available. Had we terminated the command **with(student)** with a colon, this list would not have been displayed.

Now, let's enter a function beneath which we want the area, and let's invoke the **leftbox** command.

- `f := x^2 + 1;`

$$f := x^2 + 1$$

- `leftbox(f, x = 0..1);`

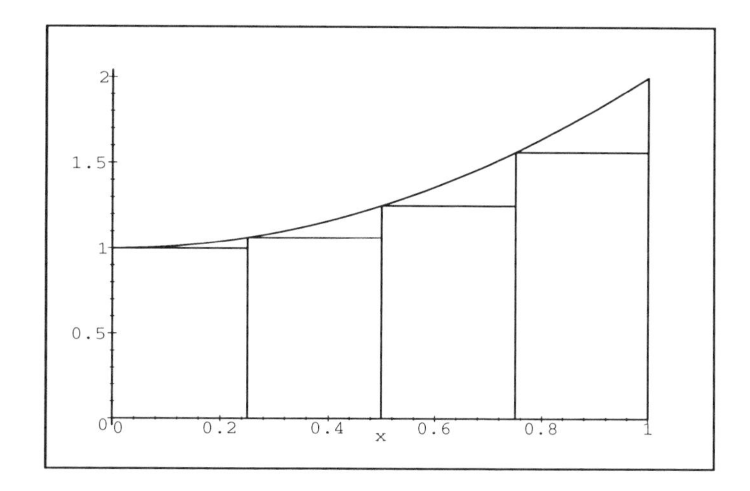

This command has created a graph of the function $f(x)$ and (by default) four rectangles approximating the area under $f(x)$. Each rectangle has its height determined by evaluating $f(x)$ at the left edge of the uniformly spaced subintervals over which the rectangle sits. The implications of "left" in the **leftbox** command are best seen in contrast to the implications of "right" in the **rightbox** command.

```
• rightbox(f, x = 0..1);
```

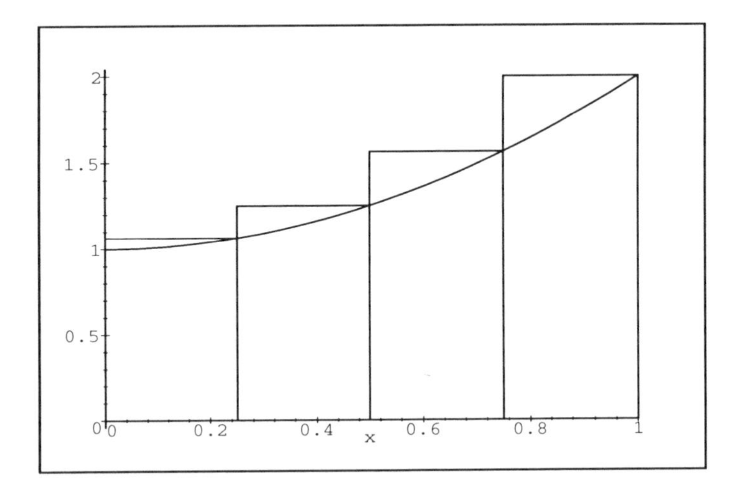

So, we have a graphic way of exploring Riemann sums. But first, how can we get past the default value of just four rectangles?

```
• leftbox(f, x = 0..1, 10);
```

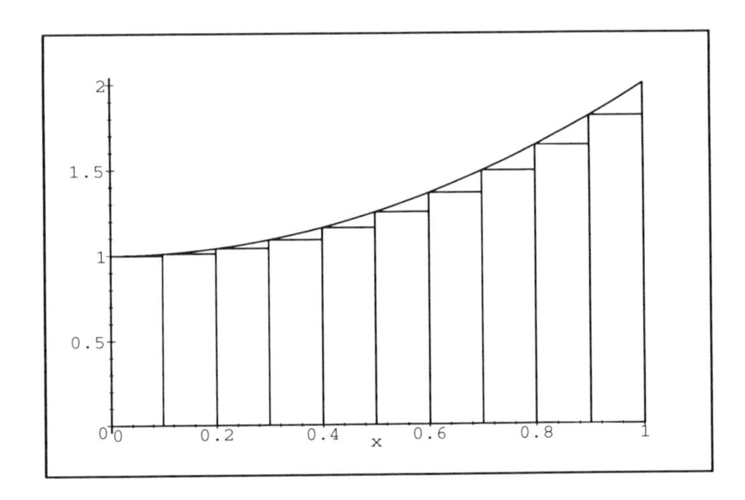

A typical reaction at this point would be the creation of graphs that show 100, 1000, and even 5000 rectangles. Thus,

- `leftbox(f, x = 0..1, 100);`

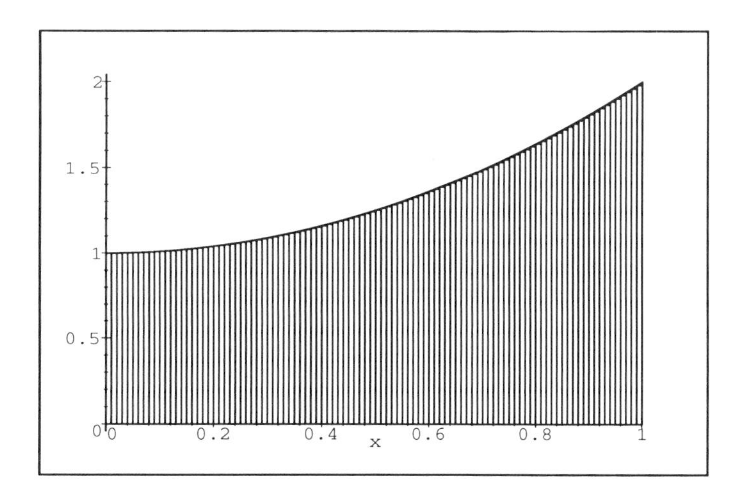

One good picture is worth a thousand words. This single picture captures the conceptual basis for the definition of the Riemann integral we seek to introduce. But in addition to these pictures, Maple has tools that help in the analytical study of the definite integral as area under a curve.

Corresponding to the **leftbox** (**rightbox**) command is the **leftsum** (**rightsum**) command that will create the analytic representation of the areas of the approximating rectangles. Thus,

- `q := leftsum(f, x = 0..1);`

$$q := \frac{1}{4}\left(\sum_{i=0}^{3}\left(\frac{1}{16}i^2+1\right)\right)$$

Maple has created the nodes $xi = a + i\frac{(b-a)}{n}$, $i = 0, \ldots, n$, where $a = 0, b = 1$, and $n = 4$, the default value for the number of subintervals. The $1/4$ in front of the summation sign is just $\frac{(b-a)}{n}$, the width of a rectangle. By looking at the graphs produced by the **leftbox** and **rightbox** commands, we readily distinguish between the $0..3$ for left-evaluation shown above and the $1..4$ for right-evaluations that we illustrate below.

- `rightsum(f, x = 0..1);`

$$\frac{1}{4}\left(\sum_{i=1}^{4}\left(\frac{1}{16}i^2+1\right)\right)$$

Next, extract a value from the **leftsum** with the default four rectangles.

- `value(q);`

$$\frac{39}{32}$$

Can we do this with more rectangles?

- `q := leftsum(f, x = 0..1, 10);`

$$q := \frac{1}{10}\left(\sum_{i=0}^{9}\left(\frac{1}{100}i^2+1\right)\right)$$

- `value(q);`

$$\frac{257}{200}$$

YES! Keep taking more rectangles using $50, 100, 500, 1000, 5000,$ etc. as the number of rectangles. Then, finally sum to an indeterminate n and take the limit as n becomes infinite.

- `q := leftsum(f, x = 0..1, n);`

$$q := \frac{\sum_{i=0}^{n-1}\left(\frac{i^2}{n^2}+1\right)}{n}$$

- `q1 := value(q);`

$$q1 := \frac{\frac{4}{3}n-\frac{1}{2}+\frac{1}{6}\frac{1}{n}}{n}$$

- `limit(q1, n = infinity);`

$$\frac{4}{3}$$

At the appropriate moment we can now define this process as "integration" and show that Maple has a built-in command that accomplishes these steps. The parallel with differentiation as a built-in operator for steps first implemented individually is obvious. The Maple command is here given by **int** .

- `int(f, x = 0..1);`

$$\frac{4}{3}$$

It is instructive to create the sums produced by the **leftsum** and **rightsum** commands. This provides a basis for a generalization to functions of several variables. An implementation of **leftsum** for the same function used above can be obtained as follows.

We first define an increment h , then $n-1$ interior partition points along the x-axis, and finally the function values at the partition points.

- `h := 1/n;`

$$h := \frac{1}{n}$$

- `xj := h*j;`

$$xj := \frac{j}{n}$$

- `fj := subs(x = xj, f);`

$$fj := \frac{j^2}{n^2} + 1$$

The approximating sum can now be obtained as an inert sum.

- `Q := Sum(fj * h, j = 0..n-1);`

$$Q := \sum_{j=0}^{n-1} \frac{\frac{j^2}{n^2} + 1}{n}$$

As before, we evaluate the inert sum via the **value** command.

- `Q1 := value(Q);`

$$Q1 := \frac{4}{3} - \frac{1}{2}\frac{1}{n} + \frac{1}{6}\frac{1}{n^2}$$

And again, the limit as n goes to infinity is

- `limit(Q1, n = infinity);`

$$\frac{4}{3}$$

It is even possible to join the inert form of the **limit** command with the inert sum.

- `QQ := Limit(Sum(fj*h, j = 0..n-1), n = infinity);`

$$QQ := \lim_{n \to \infty} \sum_{j=0}^{n-1} \frac{\frac{j^2}{n^2} + 1}{n}$$

Then, the **value** command produces the definite integral.

- `value(QQ);`

$$\frac{4}{3}$$

Some also find it useful to create the area function $A(x)$ that represents the area under $f(x)$ from a to x.

- `A := limit(value(leftsum(f, x = 0..x, n)), n = infinity);`

$$A := \frac{1}{3}\, x\, (x^2 + 3)$$

A graph is reassuring, if not helpful.

- `plot(A, x = 0..3);`

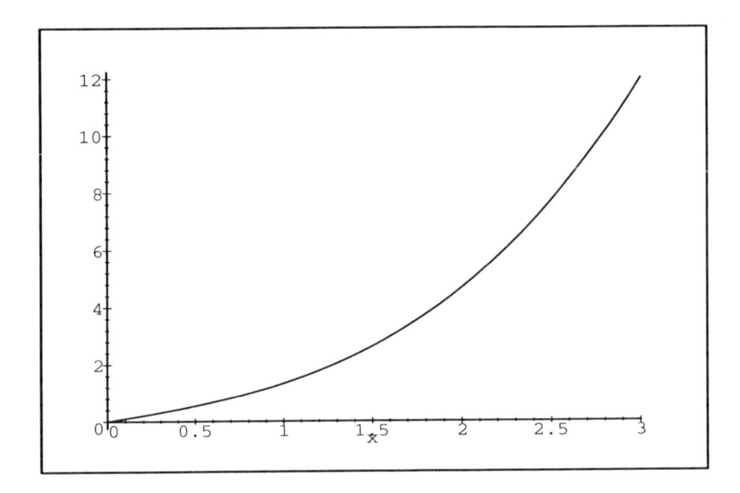

More important than the plot of the area function $A(x)$ is the function's analytic behavior.

- `diff(A, x);`

$$x^2 + 1$$

That the derivative of the area function $A(x)$ is the function under which we computed the area is remarkable. The implied connection between differentiation and integration should now be made explicit, culminating in a statement of the Fundamental Theorem of the Calculus.

Unit 13: Deriving Simpson's Rule

Maple's student package contains a built-in command for Simpson's Rule for approximate numeric integration and an exploration of the companion built-in Trapezoidal Rule appears in Unit 14. Here, we explore a derivation of Simpson's Rule.

The essence of the derivation is an interpolation. Three successive knots on a given curve are interpolated by a parabola, and the area under this parabola is computed. The resulting expression for the area under the interpolating parabola is Simpson's Rule.

Take the knots as $(a, y1)$, $(a + h, y2)$, and $(a + 2h, y3)$.

Take the interpolating parabola as

$$y = Ax^2 + Bx + C$$

Do the interpolation.

- `y := A*x^2 + B*x + C;`

$$y := A\,x^2 + B\,x + C$$

Do the *three-equations in three-unknowns* calculation, as in Units 4 and 5.

- `for k from 1 to 3 do e.k := subs(x = a + (k-1)*h, y) = y.k; od;`

$$e1 := A\,a^2 + B\,a + C = y1$$

$$e2 := A\,(\,a + h\,)^2 + B\,(\,a + h\,) + C = y2$$

$$e3 := A\,(\,a + 2\,h\,)^2 + B\,(\,a + 2\,h\,) + C = y3$$

Now solve for A, B, and C.

- `q := solve({e1,e2,e3}, {A,B,C});`

$$q := \left\{A = \frac{1}{2}\,\frac{y3 + y1 - 2\,y2}{h^2}, C = \frac{1}{2}(\right.$$
$$a^2\,y3 + a^2\,y1 - 2\,a^2\,y2 + a\,h\,y3 + 3\,a\,y1\,h - 4\,a\,y2\,h + 2\,h^2\,y1\,)\Big/h^2$$
$$\left., B = -\frac{1}{2}\,\frac{2\,a\,y3 + 2\,a\,y1 - 4\,a\,y2 + h\,y3 + 3\,h\,y1 - 4\,h\,y2}{h^2}\right\}$$

To create the actual interpolating parabola, use the substitution device.

- `yy := subs(q, y);`

$$yy := \frac{1}{2}\,\frac{(\,y3 + y1 - 2\,y2\,)\,x^2}{h^2}$$
$$-\frac{1}{2}\,\frac{(\,2\,a\,y3 + 2\,a\,y1 - 4\,a\,y2 + h\,y3 + 3\,h\,y1 - 4\,h\,y2\,)\,x}{h^2} + \frac{1}{2}($$
$$a^2\,y3 + a^2\,y1 - 2\,a^2\,y2 + a\,h\,y3 + 3\,a\,y1\,h - 4\,a\,y2\,h + 2\,h^2\,y1\,)\Big/h^2$$

Now integrate from $x = a$ to $x = a + 2h$.

The syntax of the integration command is transparent. The syntax for the indefinite integral would end at the x. The spaces are not necessary. They are there for easier reading.

- `yyy := int(yy, x = a .. a + 2*h);`

$$yyy := \frac{1}{12}(a + 2h)(2\,y3\,h^2 + 2\,h^2\,y1 - 4\,a\,y2\,h + 5\,a\,y1\,h - a\,h\,y3 + 2\,a^2\,y3 + 2\,a^2\,y1 - 4\,a^2\,y2 + 8\,y2\,h^2)\Big/h^2 - \frac{1}{12}a(3\,a\,h\,y3 + 12\,h^2\,y1 - 12\,a\,y2\,h + 9\,a\,y1\,h - 4\,a^2\,y2 + 2\,a^2\,y3 + 2\,a^2\,y1)\Big/h^2$$

This doesn't look like Simpson's Rule! Perhaps we need to simplify things.

- `y4 := simplify(yyy);`

$$y4 := \frac{1}{3}h\,(y3 + y1 + 4\,y2)$$

Any reader disturbed by Maple's failure to write the nodes in the "proper" order is welcome to try the **sort** command. This command is really written for sorting polynomials and not indexed variables, but sometimes it does manage to change the order of things to our liking.

- `sort(y4);`

$$\frac{1}{3}(y3 + 4\,y2 + y1)\,h$$

Simpson's Rule!

Unit 14: Numerical Integration

The student package contains commands for both the Trapezoidal Rule and Simpson's Rule. Let's explore how we might use the **trapezoid** command to investigate the behavior of the Trapezoidal Rule.

Begin by loading the student package.

```
• with(student):
```

Let's see how much Maple knows about the Trapezoidal Rule. Apply the command **trapezoid** to an arbitrary function $f(x)$.

```
• trapezoid(f(x), x = a..b, n);
```

$$\frac{1}{2}\,\frac{(b-a)\left(\mathrm{f}(a)+2\left(\sum_{i=1}^{n-1}\mathrm{f}\left(a+\frac{i\,(b-a)}{n}\right)\right)+\mathrm{f}(b)\right)}{n}$$

Maple knows the formula and from this formula we could extract information about the technique. Instead, let's apply the **trapezoid** command to the function $f(x) = 1/x$ and then do an error estimate on the result.

```
• f := 1/x;
```

$$f := \frac{1}{x}$$

```
• q := trapezoid(f, x = 1..2, 10);
```

$$q := \frac{3}{40} + \frac{1}{10}\left(\sum_{i=1}^{9}\frac{1}{1+\frac{1}{10}\,i}\right)$$

Extract the value of this sum via

```
• q1 := value(q);
```

$$q1 := \frac{161504821}{232792560}$$

then convert the rational number to a floating-point equivalent.

```
• evalf(q1);
```

$$.6937714032$$

Compare this answer to the exact value of the definite integral:

```
• f1 := int(f, x = 1..2);
```

$$f1 := \ln(2)$$

- `evalf(f1 - q1);`

$$-.0006242226$$

Suppose, next, we want to know the accuracy of a numeric integration for which we do not know an antiderivative. We need to use the error estimate for the Trapezoidal Rule:

$$ERROR = \frac{b-a}{12} h^2 f''(c)$$

Since the value $x = c$ where the second derivative $f''(x)$ should be evaluated is not generally knowable, this estimate is used in a "worst-case" sense:

$$|ERROR| < \frac{b-a}{12} h^2 Max\{|f''(x)|\}$$

The challenge now is to compute the maximum of $f''(x)$ over the interval $[1, 2]$.

- `f2 := diff(f,x,x);`

$$f2 := 2 \frac{1}{x^3}$$

Rather than struggle to get an analytic and exact value for the maximum of $f''(x)$, we simply estimate this maximum value from a graph.

- `plot(f2, x = 1..2);`

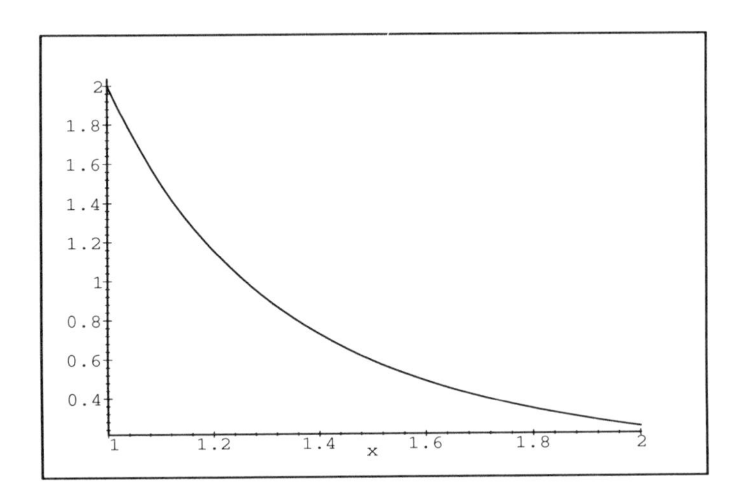

From the graph we see that the desired maximum of f "(x) occurs at $x = 1$. As noted, we could have differentiated f "(x) to locate critical points inside the interval $[a, b]$, had the graph indicated there were any.

Here, we obtain as MAX$\{f''(x)\}$ on $[1, 2]$ the value

- `m := subs(x = 1, f2);`

$$m := 2$$

Now, suppose we needed to know the value of our definite integral to an accuracy of at least $1/10,000$. This means we need an n large enough that the estimate of the error is smaller than $1/10,000$. Since $b - a = 2 - 1 = 1$, and $h = (b - a)/n = 1/n$, we get for the error estimate e

```
• e := (1/12) * (1/n)^2 * m;
```

$$e := \frac{1}{6}\frac{1}{n^2}$$

What remains is to solve the equation $e = 1/10,000$ for the smallest value of n for which the error is guaranteed to be less than $1/10,000$.

```
• v := solve(e = 1/10000, n);
```

$$v := -\frac{50}{3}\sqrt{6}, \frac{50}{3}\sqrt{6}$$

Clearly, we want the positive value, and we need to "round it up" to the next highest integer.

```
• evalf(v[2]);
```

$$40.82482906$$

If $n = 41$ in the Trapezoidal Rule, we are guaranteed an approximation to the definite integral that is accurate to at least $1/10,000$. Let's try this.

```
• q := trapezoid(f, x = 1..2, 41);
```

$$q := \frac{3}{164} + \frac{1}{41}\left(\sum_{i=1}^{40} \frac{1}{1 + \frac{1}{41} i}\right)$$

```
• q1 := value(q);
```

$$q1 := \frac{1226326031262800518064310704605583}{1769119595716035431543060307579840}$$

```
• evalf(q1 - f1);
```

$$.0000371774$$

Since this number is $3.7\ 10^{-5}$, we have computed the definite integral to the desired tolerance. It is interesting to see if there is a smaller value of n for which the desired tolerance is reached. This would happen if our estimate of MAX$\{|f''(x)|\}$ were considerably larger than the value of $f''(c)$. We simply build a table of values produced by the Trapezoidal Rule as compared to the exact value.

```
• for k from 22 to 26 do
      k, evalf(f1 - value(trapezoid(f, x = 1..2, k)));
  od;
```

$$22, -.0001290989$$

$$23, -.0001181195$$

$$24, -.0001084834$$

$$25, -.0000999800$$

$$26, -.0000924385$$

So, the Trapezoidal Rule obtains the definite integral to within the desired tolerance with $n = 25$. Such is the nature of error estimates based on Taylor expansions.

Finally, let's look at the built-in Maple facility for integrating numerically. Essentially, the **evalf** command applied to an integral tells Maple to initiate a numeric integration. There are two ways this might happen.

First, Maple might be asked to perform an integration for which it cannot supply an exact answer. Maple will return the integral unevaluated. Here is an example.

- `r := int(sin(x^2)/sqrt(1+x^3), x=0..1);`

$$r := \int_0^1 \frac{\sin(x^2)}{\sqrt{1+x^3}}\,dx$$

When Maple cannot evaluate an integral, it returns the integral in its unevaluated form. Maple will perform a numeric integration via

- `evalf(r);`

$$.2583127546$$

If greater accuracy were required you could call **evalf** with the option for more digits.

- `evalf(r, 20);`

$$.25831275464979652238$$

A second way numeric integration might be initiated is with the recognition that there is no known antiderivative expressible in terms of simple functions for the integral. The integral r is an elliptic integral, and had we realized that, we could have saved the time Maple spent hunting for an "exact value" for it. We could have entered the integral *directly* as an unevaluated integral.

- `r1 := Int(sin(x^2)/sqrt(2+x^3), x=0..1);`

$$r1 := \int_0^1 \frac{\sin(x^2)}{\sqrt{2+x^3}}\,dx$$

This saves time. Now, invoke numeric integration via the **evalf** command.

- `evalf(r1);`

$$.1978099673$$

It is usually interesting to compare the results of a numeric integration via the Trapezoidal Rule with the return of Maple's built-in numeric integrator. The comparisons are most useful when there is no exact value for the integral against which to compare.

Similar explorations for Simpson's Rule are also possible. However, that is left as an exercise.

Unit 15: Improper Integrals

Calculus II usually has a unit on improper integrals. Let's see how much help Maple is for examining improper integrals. First, let's explore an integral that can be easily misunderstood. In fact, the following integral is called an improper integral of the second kind in the calculus literature because the integrand becomes unbounded on the interior of the interval of integration.

- `int(1/x, x = -1..1);`

$$\int_{-1}^{1} \frac{1}{x}\, dx$$

Maple does not perform the integration because the integral is improper. There is a non-integrable singularity at $x = 0$, and by default, Maple does not compute the Cauchy Principal Value (CPV), which here would be zero. If Maple finds that the integral fails to exist, it will not return a value.

However, the integration command permits an optional request for the Cauchy Principal Value.

- `int(1/x, x = -1..1, 'CauchyPrincipalValue');`

$$0$$

To explain how Riemann integration handles the singularity at $x = 0$ and what a CPV is, we would like to explore this integral as it might appear in a typical calculus text. Unfortunately, Maple does not integrate $1/x$ to $\ln(|x|)$ the way a calculus text would. This makes an investigation of our integral a little more delicate.

We begin by illustrating the limiting process needed for the singularity at $x = 0$ by restricting our attention to the interval $[0, 1]$, in which x is positive and the absolute value needed in $\ln(|x|)$ is superfluous.

Let the parameter z be in the interval $[0, 1]$, and compute

- `q := int(1/x, x = z..1);`

$$q := -\ln(z)$$

Next, let z approach zero from the right so that z remains positive.

- `limit(q, z = 0, right);`

$$\infty$$

Thus, the integral $\int_0^1 \frac{1}{x} dx$ does not exist. It is no wonder, then, that the original integral returned unevaluated!

But let's see if we can perform the equivalent analysis in the interval $[-1, 0]$, where the absolute value in $\ln(|x|)$ would be significant. If we explicitly

- `assume(Z<0);`

then the integral of $1/x$ on the interval $[-1, Z]$ is given by

```
• q1 := int(1/x, x = -1..Z);
```

$$q1 := \ln(-Z\tilde{\ })$$

The tilde after the letter Z reminds us that the result in $q1$ is valid only for $Z < 0$, a result equivalent to using $\ln(|x|)$ in the integration. The limit as Z aproaches the origin from the left is then given by

```
• limit(q1, Z = 0, left);
```

$$-\infty$$

The recognition that the "values" of the integral on either side of the singularity somehow "cancel" gives rise to the notion of the Cauchy Principal Value wherein the limit in z and the limit in Z are "connected." In fact, if the rates at which z and Z approach the origin are the same, there is a "continuing cancellation" that justifies our declaring the CPV is zero.

Treat the parameter z as positive, and integrate $1/x$ on the intervals $[-1, -z]$ and $[z, 1]$. Then take the limit as z approaches zero through positive values (i.e., from the right) in a process that defines the Cauchy Principal Value.

```
• Limit(Int(1/x, x = -1..-z) + Int(1/x, x = z..1), z = 0, right);
```

$$\lim_{z \to 0+} \int_{-1}^{-z} \frac{1}{x}\,dx + \int_{z}^{1} \frac{1}{x}\,dx$$

Since Maple's integration command will not produce $\ln(|x|)$, we implement the above limit via

```
• f := ln(abs(x));
```

$$f := \ln(|x|)$$

```
• q3 := subs(x = -z, f) - subs(x = -1, f) + subs(x = 1, f)
        - subs(x = z, f);
```

$$q3 := \ln(|-z|) - \ln(|-1|) + \ln(|1|) - \ln(|z|)$$

```
• q4 := value(q3);
```

$$q4 := 0$$

Because of the symmetry involved, the argument of the limit in the CPV is here identically zero for every appropriate value of z. The limit is then trivially zero.

Next, we consider an improper integral of the first kind. Define the integrand by

```
• f := 1/(1+x^2);
```

$$f := \frac{1}{1+x^2}$$

and consider the integral

- `q5 := Int(f, x = 0..infinity);`

$$q5 := \int_0^\infty \frac{1}{1+x^2}\,dx$$

which Maple evaluates exactly via

- `value(q5);`

$$\frac{1}{2}\pi$$

The correct way to evaluate such an improper integral is to integrate to a finite upper limit, say z, and then to let z approach infinity in a limiting process. Thus

- `F := Int(f, x = 0..z);`

$$F := \int_0^z \frac{1}{1+x^2}\,dx$$

defines F as a "function" of the upper limit z. As z approaches infinity, the value of F should approach $\pi/2$. In this example Maple can compute F explicitly as a function of z.

- `F := value(F);`

$$F := \arctan(z)$$

A plot of the "function" $F(z)$ will show that it is reasonable to expect there to be a limiting value to the improper integral if z approaches infinity.

- `plot(F, z = 0..10);`

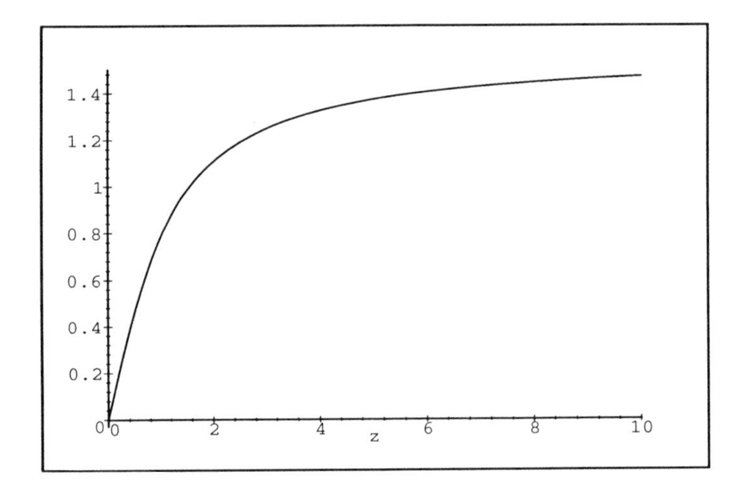

Without doubt, the limit is then $\pi/2$.

- `limit(F, z = infinity);`

$$\frac{1}{2}\pi$$

What happens when we deal with integrands for which the antiderivative is not an elementary function? How much help will Maple be in that case?

Consider the following improper integral of the first kind.

```
• g := Int(sin(x)/(1+x^3), x = 0..infinity);
```

$$g := \int_0^\infty \frac{\sin(x)}{1+x^3}\,dx$$

This integral should be written with a finite upper limit in anticipation that a limit will be taken after integrating.

```
• G := Int(sin(x)/(1+x^3), x = 0..z);
```

$$G := \int_0^z \frac{\sin(x)}{1+x^3}\,dx$$

Now what we really want to do is to make G a function of z and perhaps plot $G(z)$ against z. From such a plot we might then see what the behavior of this improper integral actually is.

The magic that causes Maple to treat G as a function of z is

```
• GG := unapply(G, z);
```

$$GG := z \to \int_0^z \frac{\sin(x)}{1+x^3}\,dx$$

The arrow notation is another way Maple has of dealing with functional relationships.

```
• plot(GG, 0..10);
```

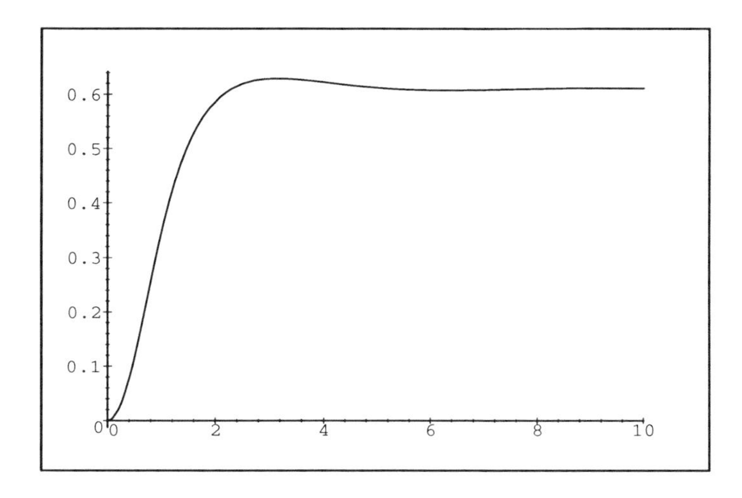

The graph suggests that for large z the integral G tends to a limit, and that the limit is on the order of about 0.6. In fact, we can evaluate G for several successively larger values of z to get a better estimate of the limit of the function $GG(z)$ as z gets large.

As an aside, notice that the syntax for plotting a function differs slightly from that for plotting an expression. The function uses a range containing no equation: the usual equation $z = 0..10$ is abridged to just the range, $0..10$.

```
• evalf(GG(10));
```

$$.6118100029$$

```
• evalf(GG(50));
```

$$.6109052369$$

```
• evalf(GG(100));
```

$$.6109119489$$

```
• evalf(GG(200));
```

$$.6109127358$$

The oscillations in these values are attributable to the $\sin(x)$ in the numerator in the integrand. Although the integral converges absolutely, it does oscillate.

Although these are the kinds of improper integrals we typically meet in calculus, they are not necessarily the kind of problems met afterwards. For example, in differential equations the Laplace transform $L[f]$ is defined as

$$\int_0^\infty \mathrm{f}(t)\, e^{(-s\,t)}\, dt$$

To see what kind of difficulties such an integral presents, obtain the Laplace transform of the function $f(t) = 1$.

```
• int(1*exp(-s*t),t = 0..infinity);
```

$$\lim_{t\to\infty-} -\frac{e^{(-s\,t)}}{s} + \frac{1}{s}$$

Maple failed to complete the integration, leaving the limit unevaluated. The problem is that Maple does not know the sign of s. If s is positive, the limit exists, but if s is negative, the integral diverges. A way to tell Maple that s is positive is via the **assume** facility. Thus,

```
• assume(s>0);
```

Now, do the integration again.

```
• int(1*exp(-s*t), t = 0..infinity);
```

$$\frac{1}{s\tilde{}}$$

The tilde following the s indicates that the result is true only if s is suitably restricted.

The use of the **assume** command always results in the tilde being attached to the variable about which an assumption has been made. To restore s to its "unassumed" state:

- `s := 's';`

$$s := s$$

Unit 16: Integration by Trigonometric Substitution

One of the most contentious issues in the introduction of computer algebra into the calculus classroom is that of "methods of integration." For some, the unit on methods of integration is a sacred rite of passage from the ignorance of the laity into the exhalted state of high-priesthood. For others, the concept of *a change of variables* in an integral is sufficient, as long as there are good tables of integrals or powerful computer algebra systems available.

Here is one way to use Maple for numbing the pain of *integration by trigonometric substitution*. We explore the idea of a change of variables in an integral, keeping in mind the tenet that rarely, if ever, will a trigonometric substitution actually be used to evaluate an integral of interest. (Either a table of integrals or a computer algebra system will be used.)

Consider, for example, the integrand

- `f := x^2 / sqrt(9 - x^2);`

$$f := \frac{x^2}{\sqrt{9 - x^2}}$$

and the indefinite integral

- `g := Int(f, x);`

$$g := \int \frac{x^2}{\sqrt{9 - x^2}}\, dx$$

It is in the integral called g that the trigonometric substitution $x = 3\sin(t)$ is to be used. What can Maple do about this?

- `with(student):`

Load the student package to access the **changevar** command. Then enter the desired change of variables in the form of a defining equation.

- `q := x = 3*sin(t);`

$$q := x = 3\sin(t)$$

Invoke the change of variables command **changevar** found in the student package.

- `g1 := changevar(q, g, t);`

$$g1 := \int 27\, \frac{\sin(t)^2 \cos(t)}{\sqrt{9 - 9\sin(t)^2}}\, dt$$

The tag $g1$ points to the transformed version of the integral g, but simplification is needed.

- `g2 := simplify(g1);`

$$g2 := 9 \int \mathrm{csgn}(\cos(t)) - \mathrm{csgn}(\cos(t))\cos(t)^2\, dt$$

While it might look like Maple has produced a mess, remember that it now knows that the square root of x^2 is not x. When "x" is a trigonometric function we really have an implicit restriction on the variable t for any "simple" form for the integrand.

There is an "override" option to **simplify** that tells Maple to **simplify** the square root of x^2 without regard to signs.

- `g3 := simplify(g1, symbolic);`

$$g3 := 9 \int -\cos(t)^2 + 1\,dt$$

Often, a computation like this is carried out "formally" with an implicit realization that t must be restricted in some fashion, with that restriction being articulated after the formal result has been obtained.

Typical comments at this point usually involve Maple's not changing the integrand in $g3$ to $\sin^2(t)$. (Perhaps Maple's failure to use parentheses around the integrand in $g3$ is the more curious behavior.)

To get Maple to evaluate the integral:

- `g4 := value(g3);`

$$g4 := \frac{9}{2}t - \frac{9}{2}\cos(t)\sin(t)$$

Clearly, we need to return to x. Solve q, the original transformation equation $x = 3sin(t)$, for t so that we have $t = t(x)$ that can be substituted into each t in $g4$.

- `T := solve(q, t);`

$$T := \arcsin\left(\frac{1}{3}x\right)$$

Please notice we avoid using the letter "t" as a tag for the solution of equation q. The "rule" is never to use as a name on the left of the assignment operator (:=) a working variable in use on the right.

The substitution $t = t(x)$ creates problems in trigonometry that should not be confused with the essence of the calculus.

- `g5 := subs(t = T, g4);`

$$g5 := \frac{9}{2}\arcsin\left(\frac{1}{3}x\right) - \frac{9}{2}\cos\left(\arcsin\left(\frac{1}{3}x\right)\right)\sin\left(\arcsin\left(\frac{1}{3}x\right)\right)$$

The notion of a change of variables in an integral becomes hopelessly entangled with the details of trigonometry if the following simplification cannot be carried out painlessly.

- `simplify(g5);`

$$\frac{9}{2}\arcsin\left(\frac{1}{3}x\right) - \frac{1}{2}\sqrt{9 - x^2}\,x$$

How does this compare to Maple's direct integration of the integrand f?

- `int(f, x);`

$$\frac{9}{2}\arcsin\left(\frac{1}{3}x\right) - \frac{1}{2}\sqrt{9 - x^2}\,x$$

Ah! Perfect agreement! By consigning to Maple the manipulative details stemming from the trigonometry we were able to illustrate the meaning of changing variables in an integral. It is this "resequencing of skills" that Maple makes possible in the calculus.

Unit 17: Integration by Parts

The companion of *Integration by Trigonometric Substitution* is *Integration by Parts*. Let's explore one way Maple can be used to learn about integration by parts. This approach is predicated on the belief that real integrals are never actually done "by parts" because in real life such integrals are found in tables of integrals.

Don't misunderstand. There is a real need for mastering the concept of integration by parts. Specific integrals are rarely evaluated by a technique such as parts integration. Typically, a table of integrals or a computer algebra system is used. But integration by parts is extremely important as a theoretical tool in such applications as the Laplace transform, asymptotic expansions of integrals, and weak solutions of differential equations. Rote practice in drill exercises is not the only way to gain the expertise needed with parts integraton.

We illustrate this claim by using the **intparts** command in Maple's student package.

- `with(student):`

Consider a typical drill exercise in the realm of parts integration.

- `q := Int(x*sin(x), x);`

$$q := \int x \sin(x)\, dx$$

Invoke Maple's **intparts** command which carries out parts integration, provided we designate the factor deemed to be "u" in an integrand of the form udv.

- `q1 := intparts(q, x);`

$$q1 := -x \cos(x) - \int -\cos(x)\, dx$$

The syntax calls for a declaration of which part of the integrand udv is to be taken as u . As you can see, Maple returns the equivalent of "uv minus the integral of vdu." In a classroom setting this result should be fully analyzed to identify each of u, v, du , and dv. Thus, one gains experience with parts integration without actually viewing it as a tool for evaluating specific integrals.

To get Maple to complete the integration, use the **value** command from the student package.

- `value(q1);`

$$-x \cos(x) + \sin(x)$$

Further insight into parts integration comes when an alternative choice for u is made.

- `q2 := intparts(q, sin(x));`

$$q2 := \frac{1}{2} \sin(x)\, x^2 - \int \frac{1}{2} \cos(x)\, x^2\, dx$$

Incidentally, $q2$ yields the same result as $q1$.

- `value(q2);`

$$-x\cos(x) + \sin(x)$$

Again, identifying u , v, du, and dv is essential in the learning process. Further, comparing the integrands in $q1$ and $q2$ leads to the "optimization" question: which choice of u leads to the "simplest" new integral? (One wag has been known to respond that *all* choices are equivalent because Maple evaluates any of the new integrals!)

Unit 18: Integration by Parts Twice

It is very tempting to use Maple to implement the "integrate by parts twice and solve for the unknown integral" technique used for evaluating that classic integral

$$\int e^{ax} \sin(bx)\, dx$$

Typically, when the realization first hits that Maple should be able to do the repeated integration by parts, the focus becomes "can I be clever enough to get Maple to do this" or "is Maple powerful enough to do it."

We first demonstrate that Maple will indeed reproduce this most difficult of parts integration problems in the calculus syllabus. When we are done we hope the reader concludes that not everything in the present syllabus remains appropriate in light of software tools like Maple.

Bring in the student package and define the integrand for the exercise.

- `with(student):`

Define the integral as an unevaluated integral, and call it q .

- `q := Int(exp(a*x)*sin(b*x), x);`

$$q := \int e^{(a\,x)} \sin(\,b\,x\,)\, dx$$

Now, integrate by parts twice. Let u be the trigonometric function in each case.

- `q1 := intparts(q, sin(b*x));`

$$q1 := \frac{\sin(\,b\,x\,)\, e^{(a\,x)}}{a} - \int \frac{\cos(\,b\,x\,)\, b\, e^{(a\,x)}}{a}\, dx$$

- `q2 := intparts(q1, cos(b*x));`

$$q2 := \frac{\sin(\,b\,x\,)\, e^{(a\,x)}}{a} - \frac{\cos(\,b\,x\,)\, e^{(a\,x)}\, b}{a^2} + \int - \frac{\sin(\,b\,x\,)\, b^2\, e^{(a\,x)}}{a^2}\, dx$$

The integral in $q2$ is a multiple of the original integral q. Since $q2$ is an alternate representation of the original integral q, we set up the equation $q = q2$ and solve for the unknown integral.

Give the name $q3$ to the equation $q = q2$.

- `q3 := q = q2;`

$$q3 := \int e^{(a\,x)} \sin(\,b\,x\,)\, dx =$$

$$\frac{\sin(\,b\,x\,)\, e^{(a\,x)}}{a} - \frac{\cos(\,b\,x\,)\, e^{(a\,x)}\, b}{a^2} + \int - \frac{\sin(\,b\,x\,)\, b^2\, e^{(a\,x)}}{a^2}\, dx$$

Maple will make no further progress solving this equation for the original integral because it does not see the integral on the right as being a multiple of the integral on the left. The cure is to move the constants in the integral on the right to a place in front of the integral. The **simplify** command will provide this service.

Simplify equation $q3$ and call the result $q4$.

- `q4 := simplify(q3);`

$$q4 := \int e^{(a\,x)} \sin(b\,x)\,dx = \frac{e^{(a\,x)} \sin(b\,x)\,a - \cos(b\,x)\,e^{(a\,x)}\,b - b^2 \int e^{(a\,x)} \sin(b\,x)\,dx}{a^2}$$

Now Maple will be able to combine the integrals and "solve for the unknown integral." We'll use **isolate** rather than **solve** because **isolate** writes the item being "solved for" on the left of an equation, thereby creating a complete mathematical statement.

- `q5 := isolate(q4, q);`

$$q5 := \int e^{(a\,x)} \sin(b\,x)\,dx = \frac{e^{(a\,x)} \sin(b\,x)\,a - \cos(b\,x)\,e^{(a\,x)}\,b}{a^2 + b^2}$$

Finally, we can obtain the simplified form of the righthand side via

- `factor(q5);`

$$\int e^{(a\,x)} \sin(b\,x)\,dx = \frac{e^{(a\,x)}\,(\sin(b\,x)\,a - \cos(b\,x)\,b)}{a^2 + b^2}$$

We made Maple perform the most difficult problem involving integration by parts that the typical calculus text contains. Did the exercise really illuminate the concept of integration by parts? Does integrating by parts twice and solving for the unknown integral in itself really generate insight into the nature of mathematics? We leave the reader to ponder that question while we suggest an additional parts integration activity.

Consider the following companion to the above calculation. In it we take as u the exponential function instead of the trigonometric function. The same set of steps will be carried out, and we will discover that the method still works. Perhaps it is the discovery that both alternatives work that is more important than the actual mechanics of the integrations.

Thus

- `Q1 := intparts(q,  exp(a*x));`

$$Q1 := -\frac{e^{(a\,x)} \cos(b\,x)}{b} - \int -\frac{a\,e^{(a\,x)} \cos(b\,x)}{b}\,dx$$

- `Q2 := intparts(Q1, exp(a*x));`

$$Q2 := -\frac{e^{(a\,x)} \cos(b\,x)}{b} + \frac{e^{(a\,x)} \sin(b\,x)\,a}{b^2} + \int -\frac{a^2\,e^{(a\,x)} \sin(b\,x)}{b^2}\,dx$$

- `Q3 := q = Q2;`

$$Q3 := \int e^{(a\,x)} \sin(b\,x)\,dx = -\frac{e^{(a\,x)}\cos(b\,x)}{b} + \frac{e^{(a\,x)}\sin(b\,x)\,a}{b^2} + \int -\frac{a^2\,e^{(a\,x)}\sin(b\,x)}{b^2}\,dx$$

- `Q4 := simplify(Q3);`

$$Q4 := \int e^{(a\,x)} \sin(b\,x)\,dx = -\frac{\cos(b\,x)\,e^{(a\,x)}\,b - e^{(a\,x)}\sin(b\,x)\,a + a^2 \int e^{(a\,x)}\sin(b\,x)\,dx}{b^2}$$

- `Q5 := isolate(Q4, q);`

$$Q5 := \int e^{(a\,x)} \sin(b\,x)\,dx = \frac{e^{(a\,x)}\sin(b\,x)\,a - \cos(b\,x)\,e^{(a\,x)}\,b}{a^2 + b^2}$$

Now, apply the **factor** command, and note that the integral comes out the same!

- `factor(Q5);`

$$\int e^{(a\,x)} \sin(b\,x)\,dx = \frac{e^{(a\,x)}\,(\sin(b\,x)\,a - \cos(b\,x)\,b)}{a^2 + b^2}$$

Simply having Maple reproduce the traditional curriculum might not be the most effective and efficient use of technology. Maple allows new learning activities to be implemented, and the insight produced by these new activities should be of greater importance than the manipulative skills stressed in the traditional syllabus.

Unit 19: Surface Area of a Solid of Revolution

Find the surface area generated when the triangle with vertices at $a:(1,1)$, $b:(5,9)$, and $c:(4,14)$ is rotated about the line $x=7$.

This exercise will demonstrate how Maple can lighten the load of "grunt" work for such a routine, but tedious, calculation.

First, we need a picture.

```
• plot({[[1,1],[5,9],[4,14],[1,1]], [7, y, y = 0..15]},
        x = 0..8);
```

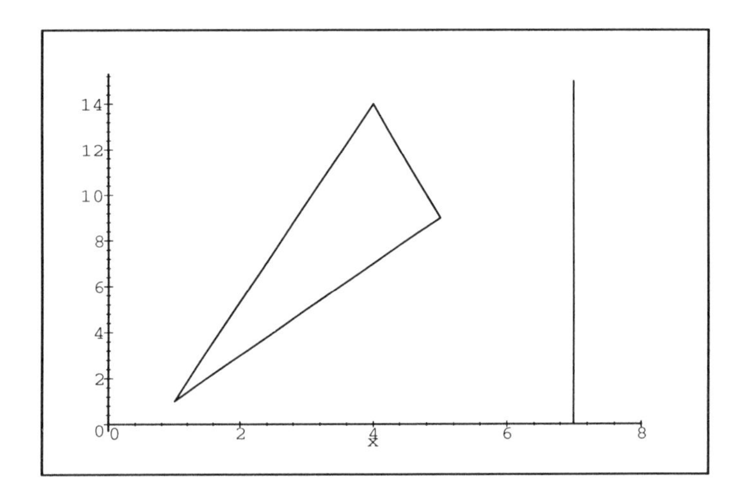

Maple "connects the dots" if you list the coordinates either as $[[x1,y1],[x2,y2],...]$, or as $[x1,y1,x2,y2,...]$.

Let's label this graph by using the **textplot** command in the plots package.

The data structures for the graph and the text to be written on the graph are superimposed via the **display** command. We access a plot data structure by assigning it to a variable, being sure to terminate the command with a colon (or else the complete plot data structure will be printed to the screen).

```
• with(plots):
  f1 := plot({[[1,1],[5,9],[4,14],[1,1]], [7, y, y = 0..15]},
             x = 0..8):
  f2 := textplot({[.75,1,'a'], [5.25,9,'b'], [4,15,'c']}):
```

- `display([f1, f2]);`

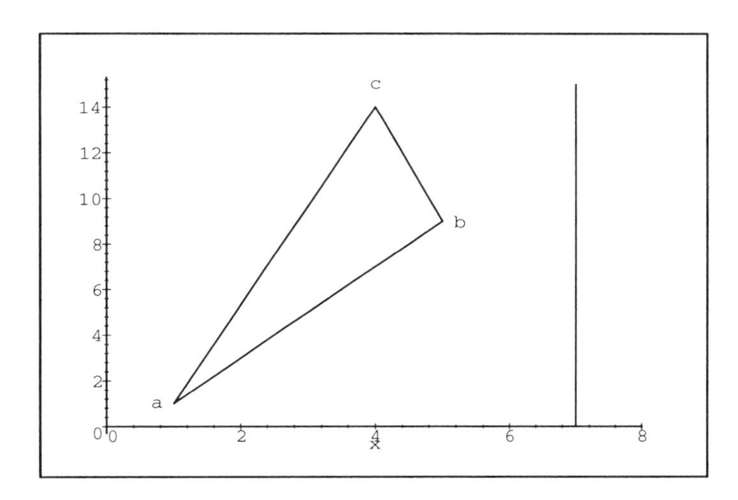

We need the equations of the lines bounding each edge of the triangle. This is a data entry and data management problem. A well-conceived plan of attack simplifies much of the work in the sequel.

Assign the letters a , b, and c to the vertices of the triangle, and have Maple calculate the slopes of the segments connecting the vertices. Load the student package for access to the **slope** command.

- ```
  with(student):
  a := [1,1];
  ```

$$a := [1, 1]$$

- `b := [5,9];`

$$b := [5, 9]$$

- `c := [4,14];`

$$c := [4, 14]$$

- `mab := slope(a,b);`

$$mab := 2$$

- `mbc := slope(b,c);`

$$mbc := -5$$

- `mca := slope(c,a);`

$$mca := \frac{13}{3}$$
  ```

Next, obtain the equations of the lines bounding the triangle.

- `yab := mab * (x - 1) + 1;`

$$yab := 2\,x - 1$$

- `ybc := mbc * (x - 5) + 9;`

$$ybc := -5\,x + 34$$

- `yca := mca * (x - 4) + 14;`

$$yca := \frac{13}{3}\,x - \frac{10}{3}$$

Given the direction of the rotation involved, we will need to express each edge functionally in the form $x = x(y)$. Hence,

- `xab := solve(yab = y, x);`

$$xab := \frac{1}{2} + \frac{1}{2}\,y$$

- `xbc := solve(ybc = y, x);`

$$xbc := \frac{34}{5} - \frac{1}{5}\,y$$

- `xca := solve(yca = y, x);`

$$xca := \frac{10}{13} + \frac{3}{13}\,y$$

The distance from the line $x = 7$ to an edge of the triangle is the next ingredient of the solution.

- `rab := 7 - xab;`

$$rab := \frac{13}{2} - \frac{1}{2}\,y$$

- `rbc := 7 - xbc;`

$$rbc := \frac{1}{5} + \frac{1}{5}\,y$$

- `rca := 7 - xca;`

$$rca := \frac{81}{13} - \frac{3}{13}\,y$$

The elements of surface area for the portions of the solid of revolution generated by each edge of the triangle can be obtained next.

- `dsab := sqrt(1 + diff(xab,y)^2);`

$$dsab := \frac{1}{2}\,\sqrt{5}$$

- `dsbc := sqrt(1 + diff(xbc,y)^2);`

$$dsbc := \frac{1}{5}\sqrt{26}$$

- `dsca := sqrt(1 + diff(xca,y)^2);`

$$dsca := \frac{1}{13}\sqrt{178}$$

The surface areas are next, and this requires the evaluation of three integrals.

- `sab:=2*Pi*int(rab*dsab, y = 1..9);`

$$sab := 32\,\pi\,\sqrt{5}$$

- `sbc:=2*Pi*int(rbc*dsbc, y = 9..14);`

$$sbc := 5\,\pi\,\sqrt{26}$$

- `sca:=2*Pi*int(rca*dsca, y = 1..14);`

$$sca := 9\,\pi\,\sqrt{178}$$

Clearly, the total surface area is the sum of each of the above integrals.

- `s := sab + sbc + sca;`

$$s := 32\,\pi\,\sqrt{5} + 5\,\pi\,\sqrt{26} + 9\,\pi\,\sqrt{178}$$

As a floating-point number we get

- `evalf(s);`

$$682.1159470$$

Maple was used in a straightforward fashion to organize the calculations. Perhaps the clarity of such structuring is a significant advantage in itself.

Unit 20: A Separable Differential Equation

The following problem comes from the section *Separable Differential Equations* in a traditional calculus text. It requires no more than a separation of variables and an "integration of both sides" to produce the solution of the differential equation in an implicit form. However, we show that additional insight is possible if we ask questions like "How does the solution behave?" and "What is the nature of the implicit function describing the solution?" The role of Maple in answering these questions is essential.

The differential equation is

```
• q := diff(y(t), t) = t^2/(1 + 3*y(t)^2);
```

$$q := \frac{\partial}{\partial t}\,\mathrm{y}(t) = \frac{t^2}{1+3\,\mathrm{y}(t)^2}$$

and its solution, either in Maple or by separating variables and integrating both sides, is

```
• q1 := dsolve(q, y(t));
```

$$q1 := \mathrm{y}(t) + \mathrm{y}(t)^3 - \frac{1}{3}\,t^3 = \_C1$$

Let's replace the arbitrary constant of integration $\_C1$ that Maple has introduced with the simpler-to-type constant c.

```
• q2 := subs(_C1 = c, q1);
```

$$q2 := \mathrm{y}(t) + \mathrm{y}(t)^3 - \frac{1}{3}\,t^3 = c$$

Now, let's solve for $y(t)$ in an effort to obtain an explicit represention of the solution.

```
• q3 := solve(q2, y(t));
```

$$q3 := \%1^{1/3} - \frac{1}{3}\,\frac{1}{\%1^{1/3}},\ -\frac{1}{2}\,\%1^{1/3} + \frac{1}{6}\,\frac{1}{\%1^{1/3}} + \frac{1}{2}\,I\,\sqrt{3}\left(\%1^{1/3} + \frac{1}{3}\,\frac{1}{\%1^{1/3}}\right),$$
$$-\frac{1}{2}\,\%1^{1/3} + \frac{1}{6}\,\frac{1}{\%1^{1/3}} - \frac{1}{2}\,I\,\sqrt{3}\left(\%1^{1/3} + \frac{1}{3}\,\frac{1}{\%1^{1/3}}\right)$$
$$\%1 := \frac{1}{6}\,t^3 + \frac{1}{2}\,c + \frac{1}{18}\,\sqrt{12 + 9\,t^6 + 54\,t^3\,c + 81\,c^2}$$

A careful inspection of this expression sequence will reveal that only the first member appears to be real. There are three branches of this cubic-in-y , and two branches seem to be complex conjugates.

Grab the first (real) branch and label it y. As the need to conserve space is relieved, Maple will expand out the represention of this branch.

```
• y := q3[1];
```

$$y := \left(\frac{1}{6}\,t^3 + \frac{1}{2}\,c + \frac{1}{18}\,\sqrt{12 + 9\,t^6 + 54\,t^3\,c + 81\,c^2}\right)^{1/3}$$

$$- \frac{1}{3} \frac{1}{\left(\frac{1}{6}t^3 + \frac{1}{2}c + \frac{1}{18}\sqrt{12 + 9t^6 + 54t^3c + 81c^2}\right)^{1/3}}$$

To see what this real solution looks like, let's give the constant c some values and plot the resulting functions. The values $c = 0, 2, 10, 30$, and 68 generate solutions $y(t)$ that satisfy $y(0) = 0, 1, 2, 3$, and 4, respectively.

- `C := [0, 2, 10, 30, 68];`

$$C := [0, 2, 10, 30, 68]$$

Now we can use a Maple loop to create the five corresponding solutions, labeled $y1$ through $y5$.

- `for k from 1 to 5 do y.k := subs(c = C[k], y); od;`

$$y1 := \left(\frac{1}{6}t^3 + \frac{1}{18}\sqrt{12 + 9t^6}\right)^{1/3} - \frac{1}{3} \frac{1}{\left(\frac{1}{6}t^3 + \frac{1}{18}\sqrt{12 + 9t^6}\right)^{1/3}}$$

$$y2 := \left(\frac{1}{6}t^3 + 1 + \frac{1}{18}\sqrt{336 + 9t^6 + 108t^3}\right)^{1/3} - \frac{1}{3} \frac{1}{\left(\frac{1}{6}t^3 + 1 + \frac{1}{18}\sqrt{336 + 9t^6 + 108t^3}\right)^{1/3}}$$

$$y3 := \left(\frac{1}{6}t^3 + 5 + \frac{1}{18}\sqrt{8112 + 9t^6 + 540t^3}\right)^{1/3} - \frac{1}{3} \frac{1}{\left(\frac{1}{6}t^3 + 5 + \frac{1}{18}\sqrt{8112 + 9t^6 + 540t^3}\right)^{1/3}}$$

$$y4 := \left(\frac{1}{6}t^3 + 15 + \frac{1}{18}\sqrt{72912 + 9t^6 + 1620t^3}\right)^{1/3} - \frac{1}{3} \frac{1}{\left(\frac{1}{6}t^3 + 15 + \frac{1}{18}\sqrt{72912 + 9t^6 + 1620t^3}\right)^{1/3}}$$

$$y5 := \left(\frac{1}{6}t^3 + 34 + \frac{1}{18}\sqrt{374556 + 9t^6 + 3672t^3}\right)^{1/3} - \frac{1}{3} \frac{1}{\left(\frac{1}{6}t^3 + 34 + \frac{1}{18}\sqrt{374556 + 9t^6 + 3672t^3}\right)^{1/3}}$$

Now plot all five solutions on one set of axes.

```
• plot({y.(1..5)}, t = 0..7);
```

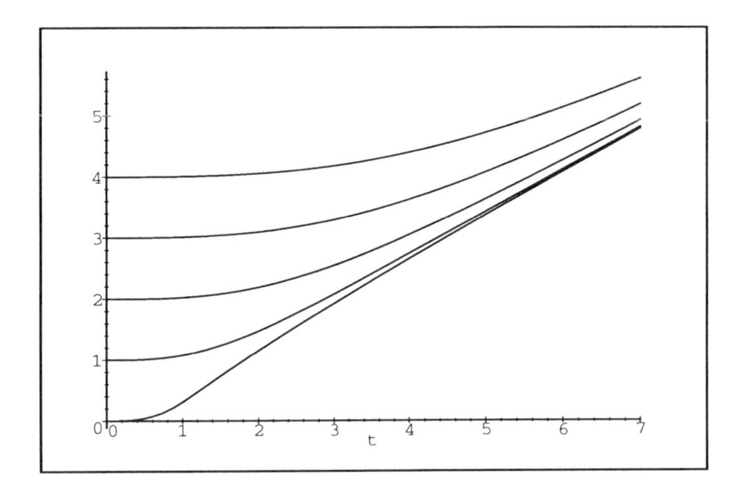

Inspection of this graph suggests that the solutions may have an oblique asymptote. To prove this, one must show that the limit of the slope (i.e., the derivative) is constant.

```
• L := limit(diff(y,t), t = infinity);
```

$$L := \frac{1}{18}\,18^{2/3}\,6^{1/3}$$

This is certainly a constant but it requires simplification.

```
• simplify(L);
```

$$\frac{1}{3}\,3^{2/3}$$

It takes a moment for this to "register," but the number is "one over the cube root of 3."

As an alternative way to see the asymptotic behavior of the solution, compute an asymptotic expansion; this is a Taylor series about the point at infinity.

```
• asympt(y,t);
```

$$\frac{1}{3}\,3^{2/3}\,t - \frac{1}{3}\,\frac{3^{1/3}}{t} + \frac{1}{3}\,\frac{3^{2/3}\,c}{t^2} + \frac{1}{3}\,\frac{3^{1/3}\,c}{t^4} + \mathrm{O}\left(\frac{1}{t^5}\right)$$

The asymptotic expansion again confirms the existence of the oblique asymptote suggested by the plot above. Clearly, introducing the asymptotic expansion in Calculus II requires rethinking the traditional syllabus, but with tools like Maple, it would certainly be possible.

We have not yet asked the really important question about this problem. A significant conceptual content of this problem resides in the study: Could the asymptotic behavior of the solution have been deduced directly fom the differential equation itself instead of from the explicit representation of the solution. Surely, that would be a mathematical inquiry of greater import than merely separating variables and integrating both sides.

Unit 21: Newton's Law of Cooling

One of the applications of integration arising in Calculus II is the separable ordinary differential equation. Separating variables and integrating both sides of the differential equation is an excuse for practicing integration. However, the assigned exercises usually involve skills far different than just the skills of integration.

In this exercise we will look at two "hard" problems that are typical of problems found in the unit on Newton's Law of Cooling. Newton's Law of Cooling embodies the assumption that heat flux is directly proportional to the temperature gradient. Somehow this message gets lost in the problem sets, because the theory appears not in the exercises but in the derivation of a formula for Newton Cooling. The practice problems are hard algebraically and often have nothing to do with the underlying ideas of Newtonian cooling. We will show how Maple minimizes the algebraic difficulties.

Problem 1:

At the spring picnic a cup of coffee is brewed. Milk is added to the coffee so that the coffee's temperature is 180 degrees F. Just then a volleyball game starts and the coffee is set on the picnic table. Seven minutes later, the temperature of the coffee is 150 degrees F; seven minutes after that, 127 degrees F. What is the temperature that spring day?

The essential ingredient of the solution is Newton's Law of Cooling:

$$T(t) = T_s + Ae^{kt}$$

T_s is the surrounding temperature, assumed to be constant. $T(t)$ is the time-varying temperature of the coffee. The constants A and k are specific to the system.

The data in Problem 1 allow us to write three equations in three unknowns.

- `e1 := 180 = Ts + A;`

$$e1 := 180 = Ts + A$$

- `e2 := 150 = Ts + A * exp(7*k);`

$$e2 := 150 = Ts + A\,e^{(7\,k)}$$

- `e3 := 127 = Ts + A * exp(14*k);`

$$e3 := 127 = Ts + A\,e^{(14\,k)}$$

Given the appropriate problem solving strategy, Maple will do the computations, as seen here.

- `solve({e1, e2, e3}, {Ts, A, k});`

$$\left\{k = \frac{1}{7}\ln\left(\frac{23}{30}\right),\ Ts = \frac{360}{7},\ A = \frac{900}{7}\right\}$$

The temperature at the spring picnic is T_s and Maple has computed that to be 360/7 (or just over 51 degrees F) - a chilly day for a picnic.

Problem 2:

At the fall picnic, on a day when the temperature was a balmy 74 degrees F, a cold drink was taken from the cooler just as the volleyball game was starting. The drink was set on the picnic table, and seven minutes later its temperature was 50 degrees F. Seven minutes after that, its temperature was 60 degrees F. How cold is the inside of the cooler?

Letting x represent the unknown initial temperature of the drink, we again get three equations in three unknowns to solve.

- `e1 := x = 74 + A;`

$$e1 := x = 74 + A$$

- `e2 := 50 = 74 + A * exp(7*k);`

$$e2 := 50 = 74 + A\,e^{(7\,k)}$$

- `e3 := 60 = 74 + A * exp(14*k);`

$$e3 := 60 = 74 + A\,e^{(14\,k)}$$

Again, the problem must be correctly formulated before Maple can help. There is no replacement for human thought and intelligence.

- `solve({e1, e2, e3}, {x, A, k});`

$$\left\{k = \frac{1}{7}\ln\left(\frac{7}{12}\right), A = \frac{-288}{7}, x = \frac{230}{7}\right\}$$

The cooler is extremely efficient, keeping drinks at 230/7 degrees F, a temperature just barely above freezing!

Unit 22: Logistic Growth

The Logistic Growth model is another example of a separable ordinary differential equation that is found in many calculus texts. The most formidable part of the discussion of this equation is the algebra of getting the solution into a "standard" form. The integrations that cause it to be included in Calculus II are not nearly as difficult as the ensuing algebra.

The equation is an example of "self-limiting" growth. The growth of $y(t)$ would be exponential except for the "self-limiting" term $1 - y/k$, k constant, that modifies the constant growth rate factor r. Hence, the population $y(t)$ satisfies the initial value problem

$$\begin{aligned} \frac{dy}{dt} &= ry(1 - y/k) \\ y(0) &= y_0 \end{aligned}$$

First, we enter the differential equation into Maple in preparation for a Maple solution of the initial value problem.

- `q := diff(y(t),t) = r*y(t)*(1 - y(t)/k);`

$$q := \frac{\partial}{\partial t} \mathrm{y}(t) = r\,\mathrm{y}(t)\left(1 - \frac{\mathrm{y}(t)}{k}\right)$$

The syntax for entering a differential equation into Maple is transparent. Note the tag q attached to the equation.

We will let Maple solve the initial value problem, again attaching a tag to Maple's return.

- `q1 := dsolve({q, y(0) = y0}, y(t));`

$$q1 := \mathrm{y}(t) = -\frac{k}{-1 - \dfrac{e^{(-r\,t)}\,(k - y0)}{y0}}$$

First, note that the response begins with $y(t) = \ldots$. This means the "stuff" on the right is not bound to the symbol $y(t)$, and unless we do something about it, we will not be able to access the "solution."

Second, Maple's response is not in the usual textbook form. We will need to do some manipulations on the solution to achieve the desired form.

Note, finally, the use of the right-hand-side command.

- `y1 := rhs(q1);`

$$y1 := -\frac{k}{-1 - \dfrac{e^{(-r\,t)}\,(k - y0)}{y0}}$$

Now that we've accessed the solution we seek to modify its form. We start by trying Maple's **simplify** command.

- `y2 := simplify(y1);`

$$y2 := -\frac{k\, y0}{-y0 - e^{(-r\,t)}\,k + e^{(-r\,t)}\,y0}$$

If we can group the two exponentials in the denominator, we will have achieved an acceptable form for the solution.

- `y3 := collect(y2, exp(-r*t));`

$$y3 := -\frac{k\, y0}{(-k + y0)\,e^{(-r\,t)} - y0}$$

In this form it is clear that if t becomes large, then y approaches k, which is therefore called the "carrying capacity" of the population.

A typical homework problem would be the following:

A 1972 population of bats in a cave is 30. Its 1977 population is 120 and its 1982 population is 150. Use the Logistic model to determine the 1994 population.

Counting time in years, starting in 1972, and taking $y0$ as 30, we have two equations in two unknowns.

- `e1 := subs(y0 = 30, t = 5, y3) = 120;`

$$e1 := -30\,\frac{k}{(-k+30)\,e^{(-5\,r)} - 30} = 120$$

- `e2 := subs(y0 = 30, t = 10,y3) = 150;`

$$e2 := -30\,\frac{k}{(-k+30)\,e^{(-10\,r)} - 30} = 150$$

- `q2 := solve({e1,e2}, {k,r});`

$$q2 := \left\{ r = \frac{1}{5}\ln(15),\, k = \frac{1680}{11} \right\}$$

- `y4 := subs(q2, y3);`

$$y4 := -\frac{1680}{11}\,\frac{y0}{\left(-\frac{1680}{11} + y0\right) e^{(-1/5\ln(15)\,t)} - y0}$$

Oops! We did not include $y0$ in the set of replacements in $q2$. We can do that after the fact, as we show next. Then, however, we'll see how we could have substituted the initial condition in the same step as we used for substituting the parameters r and k.

- `y5:= subs(y0 = 30, y4);`

$$y5 := -\frac{50400}{11}\,\frac{1}{-\frac{1350}{11}\,e^{(-1/5\ln(15)\,t)} - 30}$$

With forethought we could have made both substitutions via

- `subs(q2 union {y0 = 30}, y3);`

$$-\frac{50400}{11}\,\frac{1}{-\frac{1350}{11}\,e^{(-1/5\ln(15)\,t)}-30}$$

Thus, $q2$ is a set of equations used in the **subs** command. To add to that set, we used Maple's mathematically faithful set machinery.

A graph of the solution is certainly helpful.

- `plot(y5, t = 0..25);`

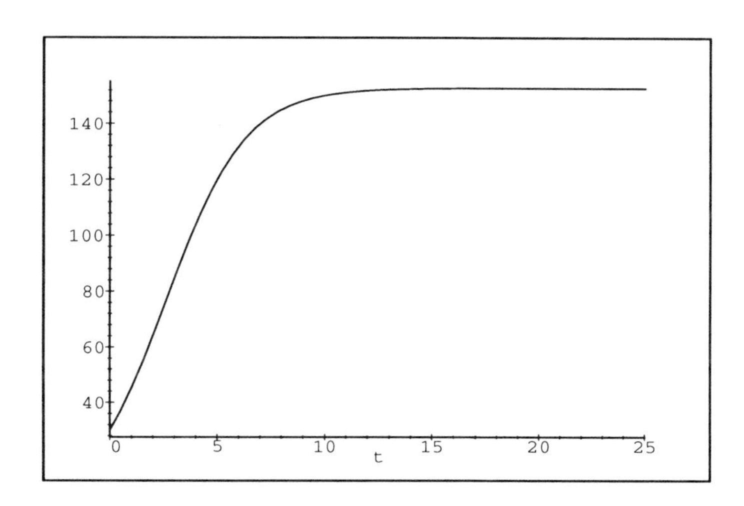

The self-limiting behavior of a population modeled by the Logistic equation is evident. The 1994 population and the carrying capacity are easily recovered.

- `p22 := subs(t = 22, y5);`

$$p22 := -\frac{50400}{11}\,\frac{1}{-\frac{1350}{11}\,e^{(-22/5\ln(15))}-30}$$

- `evalf(p22);`

$$152.7230952$$

Unit 23: L'Hôpital's Rule

Even though Maple has a very capable **limit** command, L'Hôpital's Rule is still a conceptual necessity in the calculus. Here is an investigation that combines the insight of local linearity with the operational role of the derivative in understanding what L'Hôpital's Rule actually does.

First, let's enter a function for which evaluation at $x = 0$ leads to the indeterminate form $0/0$.

- `f := -sin(x) / x;`

$$f := -\frac{\sin(x)}{x}$$

We can check Maple's response to a request to evaluate $f(0)$.

- `subs(x = 0, f);`

Error, division by zero

Why not graph $f(x)$ to see what kind of behavior we face near $x = 0$.

- `plot(f, x = -Pi..Pi);`

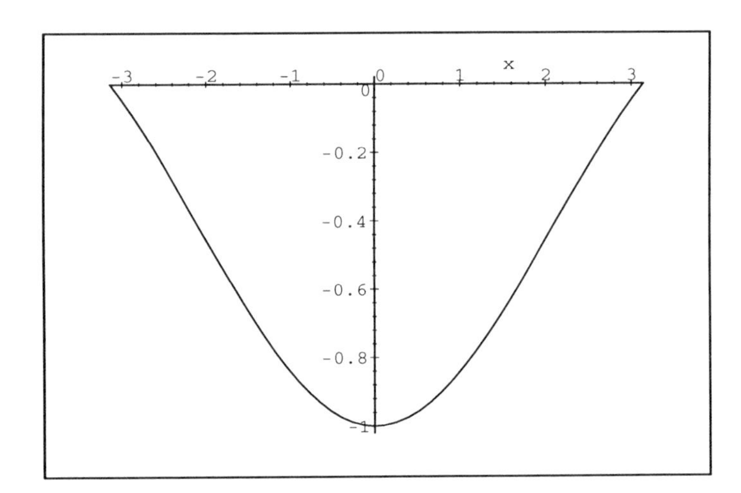

The graph shows that $f(x)$ might indeed be -1 at $x = 0$. Can we gain some insight as to why the graph of $f(x)$ shows this behavior but $f(0)$ seems to be undefined?

Perhaps we should draw graphs of the numerator and the denominator of $f(x)$.

- `plot({numer(f), denom(f)}, x = -1..1);`

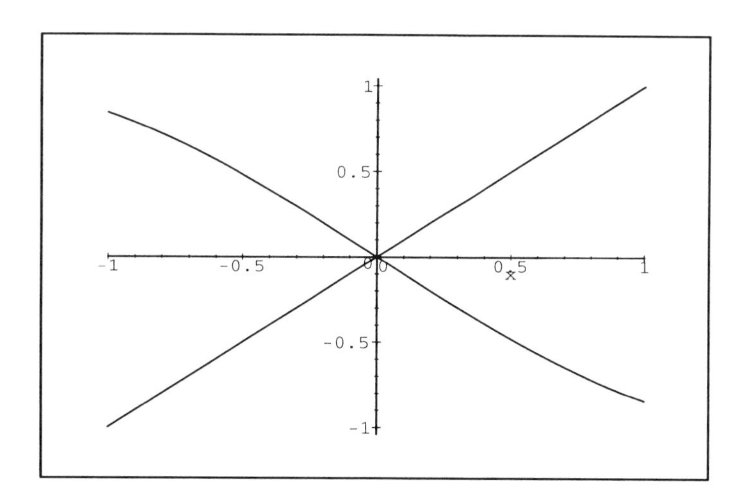

Both the numerator *and* the denominator of $f(x)$ are zero at $x = 0$. Yet the ratio seems to have the value -1 at $x = 0$. Very interesting.

Notice how this restricted (zoomed-in) view seems to look like the intersection of two lines. Perhaps there is a valuable idea hiding here. Let's create the equations of two lines that intersect at the point $(a, 0)$ on the x-axis. Let one line have slope $m1$ and the other, $m2$.

- `y1 := m1 * (x - a);`

$$y1 := m1\,(x - a)$$

- `y2 := m2 * (x - a);`

$$y2 := m2\,(x - a)$$

It would seem that the ratio $y1/y2$ is significant because it will turn out to be the ratio of slopes, $m1/m2$.

- `y1/y2;`

$$\frac{m1}{m2}$$

What would the ratio of the slopes (i.e., derivatives) be for the numerator and the denominator of $f(x)$?

- `m1 := diff(numer(f), x);`

$$m1 := -\cos(x)$$

- `m2 := diff(denom(f), x);`

$$m2 := 1$$

Ah! At $x = 0$ we see that $f(x)$ seems to behave like the ratio of the two "slopes" $-\cos(0)$ and 1, which is -1.

The way to "linearize" analytically is by examining Taylor expansions. Let's examine the ratio of the Taylor expansions of the numerator and the denominator of $f(x)$.

```
• t1 := convert(taylor(numer(f), x=0, 2), polynom);
```

$$t1 := -x$$

```
• t2 := convert(taylor(denom(f), x=0, 2), polynom);
```

$$t2 := x$$

Well, the ratio $t1/t2$ is certainly going to be -1. Hmm. Will this happen in general? Can we show this for arbitrary functions for which both the numerator and the denominator vanish at $x = a$?

Let's use $f(x)$ for the numerator and $g(x)$ for the denominator. But we will have to "unassign" f first.

```
• f := 'f';
```

$$f := f$$

Now $f(a)$ and $g(a)$ are both set to zero.

```
• f(a) := 0;
```

$$\mathrm{f}(a) := 0$$

```
• g(a) := 0;
```

$$\mathrm{g}(a) := 0$$

```
• t1 := convert(taylor(f(x), x=a, 2), polynom);
```

$$t1 := \mathrm{D}(f)(a)(x - a)$$

```
• t2 := convert(taylor(g(x), x=a, 2), polynom);
```

$$t2 := \mathrm{D}(g)(a)(x - a)$$

If we recognize Maple's notation $D(f)(a)$ as $f'(a)$, then we will immediately see that the ratio $t1/t2$ reduces to the ratio of derivatives, or slopes, at a point where both numerator and denominator have a simple zero. In fact, if we directly obtain the series expansion of the fraction $f(x)/g(x)$, we will again discover that the ratio behaves locally as the ratio of the derivatives of the numerator and the denominator.

```
• series(f(x)/g(x), x = a, 2);
```

$$\frac{\mathrm{D}(f)(a)}{\mathrm{D}(g)(a)} + \mathrm{O}(x - a)$$

Unit 24: Lines and Planes

Calculus III usually starts with a unit on lines and planes in $\mathbf{R}^3$. Let's work two standard problems from such a unit.

Problem 1:

Show that the line of intersection of the planes $x + 2y - 2z = 5$ and $5x - 2y - z = 0$ is parallel to the line $x = 2t - 3$, $y = 3t$, $z = 4t + 1$. Find the plane determined by these two lines.

One technique for obtaining the line of intersection of the two given planes is to form the cross product of the normal vectors of the planes. In support of this approach we will bring into Maple the linear algebra package that gives us access to vectors and vector operations.

- `with(linalg);`

```
Warning: new definition for   norm
Warning: new definition for   trace
```

$$[BlockDiagonal, GramSchmidt, JordanBlock, Wronskian, add, addcol, addrow, adj, adjoint, angle, augment, backsub, band, basis, bezout, blockmatrix, charmat, charpoly, col, coldim, colspace, colspan, companion, concat, cond, copyinto, crossprod, curl, definite, delcols, delrows, det, diag, diverge, dotprod, eigenvals, eigenvects, entermatrix, equal, exponential, extend, ffgausselim, fibonacci, frobenius, gausselim, gaussjord, genmatrix, grad, hadamard, hermite, hessian, hilbert, htranspose, ihermite, indexfunc, innerprod, intbasis, inverse, ismith, iszero, jacobian, jordan, kernel, laplacian, leastsqrs, linsolve, matrix, minor, minpoly, mulcol, mulrow, multiply, norm, normalize, nullspace, orthog, permanent, pivot, potential, randmatrix, randvector, rank, ratform, row, rowdim, rowspace, rowspan, rref, scalarmul, singularvals, smith, stack, submatrix, subvector, sumbasis, swapcol, swaprow, sylvester, toeplitz, trace, transpose, vandermonde, vecpotent, vectdim, vector]$$

The commands available in the linear algebra package can be found via Maple's help facility. By ending the **with(linalg)** command with the semicolon, we have caused Maple to print that list here.

Let's enter the appropriate two normal vectors and obtain their cross product.

- `n1 := array([1, 2, -2]);`

$$n1 := [1 \;\; 2 \;\; -2]$$

- `n2 := array([5, -2, -1]);`

$$n2 := [5 \ -2 \ -1]$$

Please note carefully that both $n1$ and $n2$ are "column" vectors. Even though Maple echoes them to the screen horizontally, they are indeed column vectors. With that warning issued, let's compute the required cross product.

- `n := crossprod(n1, n2);`

$$n := [-6 \ -9 \ -12]$$

A vector parallel to the given line is

- `v := array([2, 3, 4]);`

$$v := [2 \ 3 \ 4]$$

We can see by inspection that the vectors v and n are parallel. However, let's show that the cross product of these two vectors is the zero vector, again verifying that the appropriate lines are parallel.

- `crossprod(v,n);`

$$[0 \ 0 \ 0]$$

There is another way of obtaining the line of intersection of two planes. If we "solve" the two equations representing the planes for two variables in terms of the third variable, we create the parametric representation of the line of intersection.

- `e1 := x + 2*y - 2*z = 5;`

$$e1 := x + 2\,y - 2\,z = 5$$

- `e2 := 5*x - 2*y - z = 0;`

$$e2 := 5\,x - 2\,y - z = 0$$

- `solve({e1,e2}, {x,y});`

$$\left\{x = \frac{1}{2}\,z + \frac{5}{6},\, y = \frac{3}{4}\,z + \frac{25}{12}\right\}$$

If we treat z as the parameter on the line of intersection of the two planes we can write, for the direction vector of this line,

- `v1 := array([1/2, 3/4, 1]);`

$$v1 := \left[\frac{1}{2} \ \frac{3}{4} \ 1\right]$$

As before, we can compute the cross product of $v1$ with n to obtain the zero vector. Alternatively, we can multiply $v1$ by 4 to obtain v. There are two ways to multiply a vector by a scalar:

```
• scalarmul(v1, 4);
```

$$[2\ 3\ 4]$$

and the more flexible

```
• evalm(4 * v1);
```

$$[2\ 3\ 4]$$

The **evalm** (evaluate matrix) command will evaluate collections of arithmetic operations on vectors and matrices. Multiplication of two matrices requires that "&*" be used as the infix operator (to preserve order). To finish the original problem, we need to determine the equation of the plane that contains both the directions n and v. The standard way for doing this is to create a third vector "from one line to the other." The cross product of this third vector with either n or v yields the normal to the desired plane.

To create this third vector we need a point on the line of intersection and a point on the given line. A point on the given line is the obvious: $(-3, 0, 1)$. We can find a point on the line of intersection in at least two ways.

First, if the line of intersection has been found by solving for $x = x(z)$ and $y = y(z)$, then a point on this parametrically given line is the "obvious" $(5/6, 25/12, 0)$. The required third vector is now found by subtraction.

```
• p1 := array([-3, 0, 1]);
```

$$p1 := [-3\ 0\ 1]$$

```
• p2 := array([5/6, 25/12, 0]);
```

$$p2 := \left[\frac{5}{6}\ \frac{25}{12}\ 0\right]$$

```
• p := evalm(p2 - p1);
```

$$p := \left[\frac{23}{6}\ \frac{25}{12}\ -1\right]$$

The normal to the desired plane is

```
• N := crossprod(p, n);
```

$$N := [-34\ 52\ -22]$$

Let's continue to work within Maple's linear algebra package. We'll write the formula for the desired plane by using vector operations. First, a vector of variables.

```
• c := array([x, y, z]);
```

$$c := [x\ y\ z]$$

Then, the formula for the plane is of the form $ax + by + cz = d$. The lefthand side of this is

```
• s := dotprod(N, c);
```

$$s := -34\,x + 52\,y - 22\,z$$

To evaluate the constant on the righthand side, we use a known point on the plane.

```
• s1 := subs(x = p2[1], y = p2[2], z = p2[3],s);
```

$$s1 := 80$$

Consequently, the equation of the required plane is

```
• s = s1;
```

$$-34\,x + 52\,y - 22\,z = 80$$

Typically, the alternative process consists of finding a point on the line of intersection of the two planes by arbitrarily assigning one coordinate a value and simultaneously solving for the other two coordinates.

Here, let's set $z = 0$ and solve for x and y.

```
• ee1 := subs(z = 0, e1);
```

$$ee1 := x + 2\,y = 5$$

```
• ee2 := subs(z = 0, e2);
```

$$ee2 := 5\,x - 2\,y = 0$$

```
• q := solve({ee1, ee2}, {x, y});
```

$$q := \left\{x = \frac{5}{6}, y = \frac{25}{12}\right\}$$

It is no surprise that we have found the same point that we found above. Consequently, the two methods are going to generate the same solution.

Problem 2:

Show that the following two lines intersect. Find the equation of the plane determined by them.

$$\frac{x+1}{3} = \frac{y-6}{1} = \frac{z-3}{2}$$

and

$$\frac{x-6}{2} = \frac{y-11}{2} = \frac{z-3}{-1}$$

Begin by writing each line parametrically, each with a different parameter.

- `x1 := 3*t - 1;`

$$x1 := 3\,t - 1$$

- `y1 :=   t + 6;`

$$y1 := t + 6$$

- `z1 := 2*t + 3;`

$$z1 := 2\,t + 3$$

- `x2 := 2*r + 6;`

$$x2 := 2\,r + 6$$

- `y2 := 2*r + 11;`

$$y2 := 2\,r + 11$$

- `z2 :=  -r + 3;`

$$z2 := -r + 3$$

Solve the two equations $x1 = x2$ and $y1 = y2$ for t and r. Then check to see if those values of t and r cause $z1$ and $z2$ to agree.

- `q := solve({x1 = x2, y1 = y2}, {t, r});`

$$q := \{t = 1, r = -2\}$$

- `subs(q, z1);`

$$5$$

- `subs(q, z2);`

$$5$$

Ah! The lines intersect. We need the x- and y-coordinates at the point of intersection.

- `subs(q, x1);`

$$2$$

- `subs(q, y1);`

$$7$$

The point of intersection is $(2, 7, 5)$. The equation of the plane containing these two lines calls for a normal vector, obtained as the cross product of the direction vectors for each line. Since this calculation is essentially the same as the one we did in Problem 1, the completion has been left as an exercise.

Unit 25: Curvature from Every Angle

Curvature of plane and space curves is a staple of the calculus course. However, the connections between the center of curvature, the evolute and involutes, parallel curves, envelopes of families of curves, and the Bi-Fold Door Problem are not the typical fare in a calculus course. This unit explores, via Maple, curvature of plane curves - and associated concepts.

1. Introduction

Today's texts define, for any curve, the curvature $K = \left|\frac{dT}{ds}\right|$, where T is a unit tangent vector on the curve and s is arclength along the curve. For plane curves, as shown in Figure 1 below (which can be created by the accompanying Maple code), this is equivalent to $K = \frac{dw}{ds}$, where w is the angle the tangent line makes with the horizontal axis. Older texts such as the classic by Granville, Smith and Longley, define the curvature of a plane curve via $K = \frac{dw}{ds}$ and associate the sign of K with concavity. If we take a simple enough curve, we can implement this latter formula explicitly, thereby finding the curvature of a first example.

```
with(plots):
p1 := plot({[x, x^2, x = 0..2],[x, 2*x-1, x=1/4 .. 2]},
title='Figure 1'):
p2 := textplot([2/3, .2, 'w']):
display([p1,p2]);
```

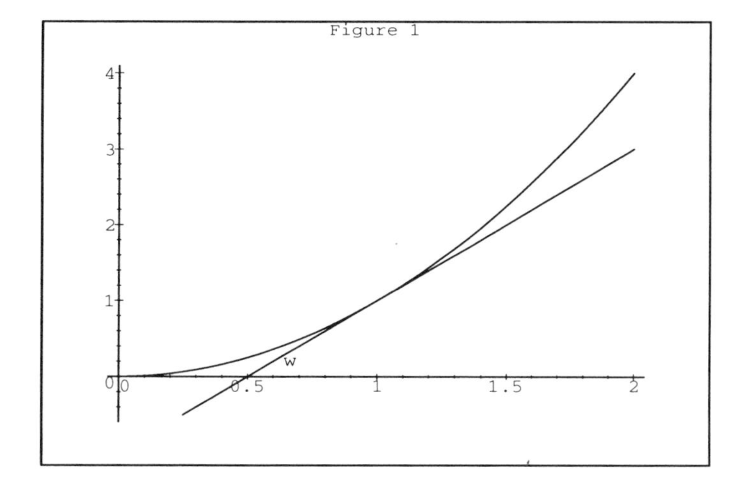

For a plane curve given as $y = y(x)$, with x in $[a, b]$, the arclength s is given by the integral

```
s = Int(sqrt(1 + diff(y(x),x)^2), x = a .. b);
```

$$s = \int_a^b \sqrt{1 + \left(\frac{\partial}{\partial x}\,y(x)\right)^2}\,dx$$

If the arclength is to be computed to a variable upper limit x, then the integral for $s = s(x)$ is now given by

- `s(x) = Int(sqrt(1 + diff(y(t),t)^2), t = a .. x);`

$$s(x) = \int_a^x \sqrt{1 + \left(\frac{\partial}{\partial t}y(t)\right)^2}\, dt$$

Let's experiment in Maple to see what some of these ideas imply. For an example, take $f(x)$ as

- `f := x^(3/2);`

$$f := x^{3/2}$$

The arclength along $f(x)$ from $x = 0$ to x is given by the integral

- `sx := int(sqrt(1 + subs(x = t, diff(f, x)^2)), t = 0..x);`

$$sx := \frac{1}{27}(4 + 9x)^{3/2} - \frac{8}{27}$$

Thus, sx is given as a function of x. To express $f(x)$ as a function of s we will have to invert $s = s(x)$ to obtain $x = x(s)$. We do this with the following algebra.

- `q := solve(sx = s, x);`

$$q := -\frac{4}{9} + \frac{1}{9}(8 + 27s)^{2/3},$$
$$-\frac{4}{9} + \frac{1}{9}\left(-\frac{1}{2}(8 + 27s)^{1/3} + \frac{1}{2}I\sqrt{3}(8 + 27s)^{1/3}\right)^2,$$
$$-\frac{4}{9} + \frac{1}{9}\left(-\frac{1}{2}(8 + 27s)^{1/3} - \frac{1}{2}I\sqrt{3}(8 + 27s)^{1/3}\right)^2$$

The first member of q is the desired real branch of the inverse $x = x(s)$. Moreover, the angle is related to the derivative through $f' = \tan(w)$. Hence, $w(s) = \arctan(f'(x(s)))$.

Hence,

- `q1 := arctan(subs(x = q[1], diff(f, x)));`

$$q1 := \arctan\left(\frac{3}{2}\sqrt{-\frac{4}{9} + \frac{1}{9}(8 + 27s)^{2/3}}\right)$$

The expression $q1$ gives $w = w(s)$ so that the differentiation defining the curvature K can be carried out explicitly.

- `q2 := diff(q1, s);`

$$q2 := 6\frac{1}{\sqrt{-\frac{4}{9} + \frac{1}{9}(8 + 27s)^{2/3}}\,(8 + 27s)}$$

This is now $K = K(s)$, which we bring into the form $K = K(x)$ by the substitution $s = s(x)$.

- `q3 := subs(s = sx, q2);`

$$q3 := 6\,\frac{1}{\sqrt{-\frac{4}{9}+\frac{1}{9}\,\left((4+9\,x)^{3/2}\right)^{2/3}}\,(4+9\,x)^{3/2}}$$

It is informative to see how to get Maple to simplify the powers in expression $q3$. Because of the correction of the "square root bug" Maple V Release 3 will not generally simplify the square root of a square. Maple needs additional information and/or guidance from the user to assure that the transformation applied will lead to correct mathematics. There are two possible approaches that work.

- `simplify(q3, power, symbolic);`

$$6\,\frac{1}{\sqrt{x}\,(4+9\,x)^{3/2}}$$

- `expand(radical(q3));`

$$6\,\frac{1}{\sqrt{x}\,(4+9\,x)^{3/2}}$$

Clearly, there are only a limited number of functions $f(x)$ for which the above calculation could be carried out in closed form. What we want is a general formula for accomplishing the same computation of the curvature.

1.1 A Formula for Curvature

Since $w = \arctan(y'(x))$ and $x = x(s)$, we need to differentiate, with respect to s , the formula $w(s) = \arctan(y'(x(s)))$. Thus,

- `q4 := arctan(D(y)(x(s)));`

$$q4 := \arctan(\mathrm{D}(y)(\mathrm{x}(s)))$$

Now, differentiate with respect to s.

- `q5 := diff(q4, s);`

$$q5 := \frac{D^{(2)}(y)(\mathrm{x}(s))\,\left(\frac{\partial}{\partial s}\,\mathrm{x}(s)\right)}{1+\mathrm{D}(y)(\mathrm{x}(s))^2}$$

The trickiest item conceptually is the equivalence of $\frac{dx(s)}{ds}$ and the reciprocal of $\frac{ds(x)}{dx}$. Computationally, it is straightforward. First, define the function $s = s(x)$.

- `q6 := s(x) = Int(sqrt(1 + D(y)(t)^2), t = a..x);`

$$q6 := \mathrm{s}(x) = \int_a^x \sqrt{1+\mathrm{D}(y)(t)^2}\,dt$$

Next, compute the derivative $\frac{ds(x)}{dx}$.

- `q7 := diff(q6, x);`

$$q7 := \frac{\partial}{\partial x} s(x) = \sqrt{1 + D(y)(x)^2}$$

Replace $\frac{dx(s)}{ds}$ with the reciprocal of $\frac{ds(x)}{dx}$ in the result for $\frac{dw}{ds}$ contained in $q5$.

- `q8 := subs(diff(x(s), s) = 1/rhs(q7), q5);`

$$q8 := \frac{D^{(2)}(y)(x(s))}{\sqrt{1 + D(y)(x)^2}\,(1 + D(y)(x(s))^2)}$$

The final simplification is the rationalization of notation for x and $x(s)$. We'll change $x(s)$ to the simpler x.

- `subs(x(s) = x, q8);`

$$\frac{D^{(2)}(y)(x)}{(1 + D(y)(x)^2)^{3/2}}$$

Typically, this result is written as

$$K = \frac{y''}{(1+(y')^2)^{3/2}}$$

The sign of the curvature will agree with that of y and hence with the concavity of $y(x)$, unless, of course, K is defined to be nonnegative.

1.2 Curvature of a Circle

No discussion of curvature is complete without the observation that the curvature of a circle is, as it should be, constant. What is not obvious is that this constant curvature is the reciprocal of the circle's radius. We can use Maple to implement the formula for the curvature K, a computation based on being able to obtain the first and second derivatives implicitly for $y(x)$ on a circle.

Begin by writing the cartesian equation for a circle with center at (h, k) and radius r.

- `q9 := (x - h)^2 + (y - k)^2 = r^2;`

$$q9 := (x - h)^2 + (y - k)^2 = r^2$$

Implicit differentiation proceeds under the assumption that y is really $y(x)$.

- `q10 := subs(y = y(x), q9);`

$$q10 := (x - h)^2 + (y(x) - k)^2 = r^2$$

Differentiation is now with respect to x throughout.

- `q11 := diff(q10, x);`

$$q11 := 2\,x - 2\,h + 2\,(\mathrm{y}(\,x\,) - k\,)\left(\frac{\partial}{\partial x}\,\mathrm{y}(\,x\,)\right) = 0$$

Solve for the first derivative $y'(x)$.

- `q12 := solve(q11, diff(y(x), x));`

$$q12 := -\frac{2\,x - 2\,h}{2\,\mathrm{y}(\,x\,) - 2\,k}$$

The extraneous 2's should probably be eliminated.

- `q13 := simplify(q12);`

$$q13 := \frac{x - h}{-\mathrm{y}(\,x\,) + k}$$

Since this is an implicit representation of the first derivative and each y in it appears as $y(x)$, we can differentiate again to obtain the second derivative.

- `q14 := diff(q13, x);`

$$q14 := \frac{1}{-\mathrm{y}(\,x\,) + k} + \frac{(\,x - h\,)\left(\frac{\partial}{\partial x}\,\mathrm{y}(\,x\,)\right)}{(\,-\mathrm{y}(\,x\,) + k\,)^2}$$

This implicit representation of the second derivative contains a first derivative which has a representation in terms of x and $y(x)$ through $q13$. Hence,

- `q15 := subs(diff(y(x), x) = q13, q14);`

$$q15 := \frac{1}{-\mathrm{y}(\,x\,) + k} + \frac{(\,x - h\,)^2}{(\,-\mathrm{y}(\,x\,) + k\,)^3}$$

We have all the ingredients needed for the curvature.

- `q16 := q15/(1 + q13^2)^(3/2);`

$$q16 := \frac{\dfrac{1}{-\mathrm{y}(\,x\,) + k} + \dfrac{(\,x - h\,)^2}{(\,-\mathrm{y}(\,x\,) + k\,)^3}}{\left(1 + \dfrac{(\,x - h\,)^2}{(\,-\mathrm{y}(\,x\,) + k\,)^2}\right)^{3/2}}$$

This expression contains square roots of squares. Since the prevailing definition of curvature does not include the absolute value, we're free to simplify, letting Maple take such square roots without absolute values. The way to tell Maple to preceed "formally" is by using the **symbolic** option in the **simplify** command.

- `q17 := simplify(q16, symbolic);`

$$q17 := \frac{1}{\sqrt{\mathrm{y}(\,x\,)^2 - 2\,\mathrm{y}(\,x\,)\,k + k^2 + x^2 - 2\,x\,h + h^2}}$$

The term under the radical sign appears to be the radius squared. To have Maple recognize that also, we simplify with respect to the side-relation defining the circle.

- `q17a := simplify(q17, {q10});`

$$q17a := \frac{1}{\sqrt{r^2}}$$

It is now anticlimactic to get Maple to produce the desired $1/r$.

- `simplify(q17a, symbolic);`

$$\frac{1}{r}$$

2. The Circle of Curvature

For each point $(x, y(x))$ on the given curve, a circle of radius $r = \frac{1}{K}$ is defined to be tangent to the curve at $(x, y(x))$. This circle is called the *Circle of Curvature*, and its center is called the *Center of Curvature*.

Momentarily, let s accept that a circle of radius R, tangent to $y(x)$ at $(x, y(x))$, is meaningful. Let's construct an example of what this might look like before we examine why such a definition might have been formulated.

Suppose we pick $y(x) = x^2$ and construct the circle of curvature at the point $(1, 1)$. The challenge is to determine where the center of curvature falls. A primitive way of doing this is to construct the equation of the normal line at $(1, 1)$ and to place the center of curvature on this line at a distance of R units from $(1, 1)$.

- `f := x^2;`

$$f := x^2$$

Next, compute the expression for $K(x)$, the curvature at any point $(x, y(x))$ on the parabola in this example.

- `K := diff(f,x,x)/(1 + diff(f,x)^2)^(3/2);`

$$K := 2\frac{1}{(1 + 4x^2)^{3/2}}$$

Since we are going to work at the point $(1, 1)$ we need to compute $K(1)$.

- `K1 := subs(x = 1, K);`

$$K1 := \frac{2}{25}\sqrt{5}$$

The *radius of curvature* for the circle of curvature we want to draw is therefore

- `R := 1/K1;`

$$R := \frac{5}{2}\sqrt{5}$$

The derivative gives the slope of the tangent line, and its negative reciprocal gives the slope of the normal line.

```
• p := subs(x = 1, diff(f, x));
```

$$p := 2$$

The equation of the normal line through $(1, 1)$ is

```
• q18 := y - 1 = -1/p*(x - 1);
```

$$q18 := y - 1 = -\frac{1}{2}x + \frac{1}{2}$$

Write this line in the form $y = ...$, being careful not to use the letter y as a name on the left.

```
• q19 := solve(q18, y);
```

$$q19 := \frac{3}{2} - \frac{1}{2}x$$

This line describes all points (x, y) that are candidates for the center of curvature. Hence, construct a circle of radius R through the fixed point $(1, 1)$ with its center on the normal line.

```
• q20 := (1 - x)^2 + (1 - q19)^2 = R^2;
```

$$q20 := (1 - x)^2 + \left(\frac{1}{2}x - \frac{1}{2}\right)^2 = \frac{125}{4}$$

The geometry tells us that, at most, two values of x can satisfy this relationship. By solving for x we will determine the location of the center of curvature.

```
• q21 := solve(q20, x);
```

$$q21 := 6, -4$$

The second of these two values puts the circle of curvature to the "left" of $(1, 1)$.

```
• xc := q21[2];
```

$$xc := -4$$

The corresponding y-coordinate is

```
• yc := subs(x = xc , q19);
```

$$yc := \frac{7}{2}$$

The parametric representation of the circle of curvature at $(1, 1)$ can now be written as

```
• u := xc + R*cos(t);
```

$$u := -4 + \frac{5}{2}\sqrt{5}\cos(t)$$

- `v := yc + R*sin(t);`

$$v := \frac{7}{2} + \frac{5}{2}\sqrt{5}\sin(t)$$

Finally, we can draw a picture of the circle of curvature, the parabola, and the normal line.

- ```
  plot({[u, v, t = 0..2*Pi], [x, f, x = -3..3],
                [x, q19, x = -4..2]}, title = 'Figure 2');
  ```

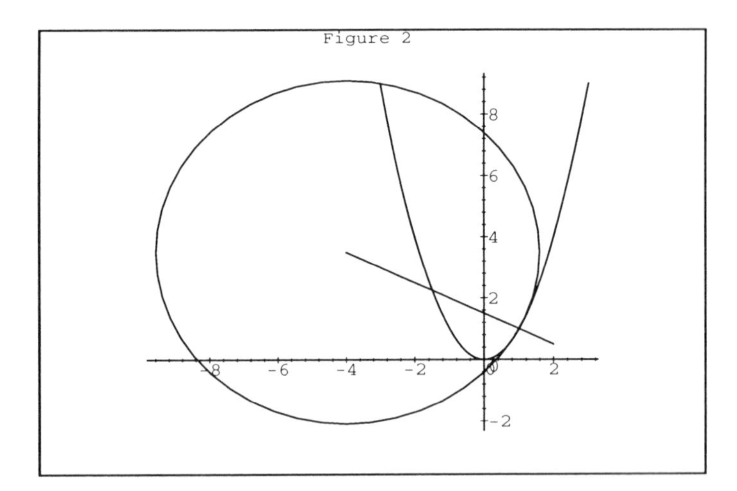

Although this is an interesting picture, there is a more pressing concern lurking. What is so special about the circle of curvature? Why is a circle with radius $1/K$ so important? What properties does this circle have?

2.1 Degree of Contact

First, let's investigate the degree of contact between the circle of curvature and the parabola. In order to avoid the complications of introducing a parameter, we'll re-write the circle of curvature in cartesian coordinates.

- `q22 := (x - xc)^2 + (y - yc)^2 = R^2;`

$$q22 := (x+4)^2 + \left(y - \frac{7}{2}\right)^2 = \frac{125}{4}$$

Let's work with an explicit representation $y = y(x)$ for this circle.

- `q23 := solve(q22, y);`

$$q23 := \frac{7}{2} + \frac{1}{2}\sqrt{61 - 4x^2 - 32x}, \frac{7}{2} - \frac{1}{2}\sqrt{61 - 4x^2 - 32x}$$

The branch that passes through the point $(1, 1)$ is the second one.
  ```

- `Y := q23[2];`

$$Y := \frac{7}{2} - \frac{1}{2}\sqrt{61 - 4\,x^2 - 32\,x}$$

This is $y = y(x)$ for the portion of the circle of curvature containing the point $(1,1)$. We wish to know both $y'(1)$ and $y''(1)$. Hence,

- `simplify(subs(x = 1, diff(Y, x)));`

$$2$$

- `simplify(subs(x = 1, diff(Y, x, x)));`

$$2$$

Just for good measure, check the third derivative.

- `simplify(subs(x = 1, diff(Y, x$3)));`

$$\frac{24}{5}$$

Well, that's a new insight into the circle of curvature. It is the circle tangent to the curve $y = y(x)$ and having second-order contact with the given curve. Do these conditions determine the circle of curvature? Is the second-degree contact property the essence of the definition of the circle of curvature?

2.2 Characterizing the Circle of Curvature

We test the hypothesis that of all circles tangent to the curve $y(x)$, the one with second-degree contact is the circle of curvature. In Section 1.2, we have already computed $y'(x)$ and $y''(x)$ for the circle with center (h,k) and radius r. In fact, these results are given in $q13$ and $q15$, respectively. Since our investigation is to take place at $(1,1)$, the first equation we need imposes the condition $y'(1) = 2$. The set braces ensure that the substitutions are made in the correct order.

- `e1 := subs({x = 1, y(x) = 1}, q13) = 2;`

$$e1 := \frac{1-h}{-1+k} = 2$$

The second equation imposes the condition $y''(1) = 2$.

- `e2 := subs({x = 1, y(x) = 1}, q15) = 2;`

$$e2 := \frac{1}{-1+k} + \frac{(1-h)^2}{(-1+k)^3} = 2$$

The expectation is that the two equations $e1$ and $e2$ determine the center of curvature.

- `q24 := solve({e1, e2}, {h, k});`

$$q24 := \left\{k = \frac{7}{2}, h = -4\right\}$$

As hypothesized, the condition of second-degree contact determines the center of curvature and, therefore, both the curvature and the radius of curvature! This latter is obtained via

- `subs(q24 union {x = 1, y = 1}, q9);`

$$\frac{125}{4} = r^2$$

We can verify this result in the general case. Let $(a, y(a))$ be an arbitrary point on the curve $y(x)$. Write ya for $y(a)$, ypa for $y'(a)$ and $yppa$ for $y''(a)$. Using the expressions for implicit derivatives computed earlier, the conditions for second-degree contact are then

- `q25 := subs({x = a, y(x) = ya}, q13) = ypa;`

$$q25 := \frac{a - h}{-ya + k} = ypa$$

and

- `q26 := subs({x = a, y(x) = ya}, q15) = yppa;`

$$q26 := \frac{1}{-ya + k} + \frac{(a - h)^2}{(-ya + k)^3} = yppa$$

Remember, our hypothesis is that the condition of second-degree contact determines the circle of curvature completely.

- `q27 := solve({q25, q26}, {h, k});`

$$q27 := \left\{k = \frac{ypa^2 + 1 + yppa\ ya}{yppa}, h = \frac{a\ yppa - ypa^3 - ypa}{yppa}\right\}$$

The closest Maple comes to putting these expressions into "standard" form is

- `q27expanded := expand(q27);`

$$q27expanded := \left\{h = a - \frac{ypa^3}{yppa} - \frac{ypa}{yppa}, k = \frac{ypa^2}{yppa} + \frac{1}{yppa} + ya\right\}$$

However, that does not diminish our accomplishment. We have characterized the center of curvature and have derived its parametric representation.

Next, let's see if we have nailed down the radius of curvature.

- `q28 := simplify(subs(q27 union {x = a, y = ya}, q9));`

$$q28 := \frac{(ypa^2 + 1)^3}{yppa^2} = r^2$$

Thus, we get for the radius of curvature

$$R = \frac{(1 + (y'(x))^2)^{3/2}}{y''(x)}$$

and for the curvature,

$$K = \frac{y''(x)}{(1+(y'(x))^2)^{3/2}}$$

In addition, we have found parametric representations for the center of curvature. Thus, as the point $(x, y(x))$ traces out the original curve, the center of curvature (h, k) will trace out another curve called the *evolute* of $y(x)$. Before we study the properties of the evolute, we pause momentarily to examine, from another angle, the characterization of the circle of curvature as the circle whose contact with the curve $y(x)$ is second-order.

2.3 Alternative Characterization of Second-Degree Contact

If a circle is passed through any three points on the curve $y(x)$ and the two "outside" points are allowed to approach the "interior" one, in the limit the circle becomes the circle of curvature. A moment's thought reveals that this is just another way of obtaining a circle with second-order contact.

It is instructive to create a realization of this insight via a specific example. Take, therefore, the curve $y = x^2$, which is presently given in Maple by f. Interpolating a circle through three points will require the general equation of a circle, which is presently $q9$ in Maple. For simplicity, we'll pick the three points $(1, y(1)), (1+m, y(1+m))$, and $(1-m, y(1-m))$.

- `e1 := subs(y = f, x = 1, q9);`

$$e1 := (1-h)^2 + (1-k)^2 = r^2$$

- `e2 := subs(y = f, x = 1 + m, q9);`

$$e2 := (1+m-h)^2 + \left((1+m)^2 - k\right)^2 = r^2$$

- `e3 := subs(y = f, x = 1 - m, q9);`

$$e3 := (1-m-h)^2 + \left((1-m)^2 - k\right)^2 = r^2$$

Solve these three equations for h, k, and r^2. (Solving for r leads to expressions containing square roots, and this complicates the mathematics needlessly.)

- `e4 := solve({e1, e2, e3},{h, k, r^2});`

$$e4 := \left\{k = \frac{7}{2} + \frac{1}{2}m^2, h = -4 + m^2, r^2 = \frac{125}{4} - \frac{15}{2}m^2 + \frac{5}{4}m^4\right\}$$

In this example we are just looking for the equation of the interpolating circle. Hence, insert the information in the set $e4$ into the equation of the circle given by $q9$.

- `e5 := subs(e4, q9);`

$$e5 := (x+4-m^2)^2 + \left(y - \frac{7}{2} - \frac{1}{2}m^2\right)^2 = \frac{125}{4} - \frac{15}{2}m^2 + \frac{5}{4}m^4$$

By inspection we can see that as m approaches zero, $e5$ approaches the circle of curvature found in Section 2. Here we propose a graphical appreciation for how $e5$ approaches the circle of curvature as m approaches zero.

Define a function that for each input value of m outputs the implicitly drawn graph of a circle in the family $e5$.

```
• e6 := t -> implicitplot(subs(m = t, e5),
     x = -10..5, y = -3..10, scaling = constrained);
```

$$e6 := t \to \text{implicitplot}(\text{subs}(m = t, e5), x = -10..5, y = -3..10, scaling = constrained)$$

Create a graph of the parabola f so that the graph can be included in each frame of a movie showing the convergence of the family $e5$ to the circle of curvature.

```
• F := plot(f, x = -10..5, y = -3..10, scaling = constrained):
```

The following Maple command creates an animation of 11 snapshots of the family $e5$ approaching the circle of curvature.

```
• display([seq(display([F,e6(1-i/10)]),i=0..10)], insequence=true);
```

It is equally interesting to see this calculation in the general case by using the three points $(a, y(a)), (a+m, y(a+m))$, and $(a-m, y(a-m))$. Of course, it results in exactly the same set of formulas for h, k and r that we found in Section 2.2.

```
• e7 := subs(x = a, y = y(a), q9);
```

$$e7 := (a-h)^2 + (y(a)-k)^2 = r^2$$

```
• e8 := subs(x = a + m, y = y(a + m), q9);
```

$$e8 := (a+m-h)^2 + (y(a+m)-k)^2 = r^2$$

```
• e9 := subs(x = a - m, y = y(a - m), q9);
```

$$e9 := (a-m-h)^2 + (y(a-m)-k)^2 = r^2$$

Again, solving for h, k and r^2 (instead of r), we get

```
• e10 := solve({e7, e8, e9}, {h, k, r^2});
```

$$e10 := \Big\{h = \frac{1}{2}\Big(2\,a\,m\,y(a+m) - 4\,a\,m\,y(a) + 2\,a\,m\,y(a-m)$$
$$- m^2\,y(a+m) + m^2\,y(a-m) - y(a+m)^2\,y(a)$$
$$+ y(a+m)^2\,y(a-m) + y(a+m)\,y(a)^2 - y(a+m)\,y(a-m)^2$$
$$- y(a)^2\,y(a-m) + y(a)\,y(a-m)^2\Big) \Big/ ($$
$$(y(a+m) - 2\,y(a) + y(a-m))\,m),$$
$$k = \frac{1}{2}\,\frac{2\,m^2 + y(a+m)^2 - 2\,y(a)^2 + y(a-m)^2}{y(a+m) - 2\,y(a) + y(a-m)},\ r^2 = \frac{1}{4}\Big($$
$$-8\,m^4\,y(a)\,y(a-m) - 2\,y(a+m)\,y(a-m)^4\,y(a)$$
$$- 8\,y(a)^3\,m^2\,y(a+m) - 8\,y(a)^3\,m^2\,y(a-m)$$

$$
\begin{aligned}
&+ 12\, m^2\, \mathrm{y}(a+m)\, \mathrm{y}(a)^2\, \mathrm{y}(a-m) - 6\, m^2\, \mathrm{y}(a+m)\, \mathrm{y}(a)\, \mathrm{y}(a-m)^2 \\
&- 6\, m^2\, \mathrm{y}(a-m)\, \mathrm{y}(a+m)^2\, \mathrm{y}(a) - 8\, m^4\, \mathrm{y}(a)\, \mathrm{y}(a+m) \\
&- 2\, m^4\, \mathrm{y}(a+m)\, \mathrm{y}(a-m) - 2\, m^2\, \mathrm{y}(a+m)^3\, \mathrm{y}(a) \\
&- 2\, m^2\, \mathrm{y}(a+m)^3\, \mathrm{y}(a-m) + 6\, m^2\, \mathrm{y}(a+m)^2\, \mathrm{y}(a)^2 \\
&+ 6\, m^2\, \mathrm{y}(a+m)^2\, \mathrm{y}(a-m)^2 - 2\, m^2\, \mathrm{y}(a-m)^3\, \mathrm{y}(a+m) \\
&+ 6\, m^2\, \mathrm{y}(a-m)^2\, \mathrm{y}(a)^2 - 2\, m^2\, \mathrm{y}(a-m)^3\, \mathrm{y}(a) \\
&- 2\, \mathrm{y}(a+m)^4\, \mathrm{y}(a)\, \mathrm{y}(a-m) + 2\, \mathrm{y}(a+m)^3\, \mathrm{y}(a)\, \mathrm{y}(a-m)^2 \\
&+ 2\, \mathrm{y}(a+m)^2\, \mathrm{y}(a)^3\, \mathrm{y}(a-m) - 6\, \mathrm{y}(a+m)^2\, \mathrm{y}(a)^2\, \mathrm{y}(a-m)^2 \\
&+ 2\, \mathrm{y}(a+m)^3\, \mathrm{y}(a-m)\, \mathrm{y}(a)^2 + 2\, \mathrm{y}(a+m)^2\, \mathrm{y}(a-m)^3\, \mathrm{y}(a) \\
&- 2\, \mathrm{y}(a+m)\, \mathrm{y}(a)^4\, \mathrm{y}(a-m) + 2\, \mathrm{y}(a+m)\, \mathrm{y}(a)^3\, \mathrm{y}(a-m)^2 \\
&+ 2\, \mathrm{y}(a+m)\, \mathrm{y}(a-m)^3\, \mathrm{y}(a)^2 + 8\, m^4\, \mathrm{y}(a)^2 + \mathrm{y}(a+m)^4\, \mathrm{y}(a)^2 \\
&- 2\, \mathrm{y}(a+m)^3\, \mathrm{y}(a)^3 + \mathrm{y}(a+m)^4\, \mathrm{y}(a-m)^2 \\
&- 2\, \mathrm{y}(a+m)^3\, \mathrm{y}(a-m)^3 + 5\, m^4\, \mathrm{y}(a+m)^2 + 5\, m^4\, \mathrm{y}(a-m)^2 \\
&+ m^2\, \mathrm{y}(a+m)^4 + m^2\, \mathrm{y}(a-m)^4 + 4\, \mathrm{y}(a)^4\, m^2 + \mathrm{y}(a+m)^2\, \mathrm{y}(a)^4 \\
&+ \mathrm{y}(a+m)^2\, \mathrm{y}(a-m)^4 + \mathrm{y}(a)^4\, \mathrm{y}(a-m)^2 - 2\, \mathrm{y}(a)^3\, \mathrm{y}(a-m)^3 \\
&+ \mathrm{y}(a)^2\, \mathrm{y}(a-m)^4 + 4\, m^6\Big) \Big/ \Big((\mathrm{y}(a+m) - 2\, \mathrm{y}(a) + \mathrm{y}(a-m))^2\, m^2 \\
&\Big)\Big\}
\end{aligned}
$$

This time we are interested in the limiting behaviors of h, k and ultimately r. We therefore extract from the set $e10$ the individual results $h(m)$, $k(m)$ and $r(m)^2$. Since we can already see the complexity of these expressions, we'll use a colon to suppress repeating this printout.

```
• hh := subs(e10, h):
  kk := subs(e10, k):
  rr := subs(e10, r^2):
```

Taking the limits as m approaches 0 we find

```
• H := limit(hh, m = 0);
```

$$H := \frac{-\mathrm{D}(y)(a) + a\, D^{(2)}(y)(a) - \mathrm{D}(y)(a)^3}{D^{(2)}(y)(a)}$$

```
• K := limit(kk, m = 0);
```

$$K := \frac{1 + \mathrm{y}(a)\, D^{(2)}(y)(a) + \mathrm{D}(y)(a)^2}{D^{(2)}(y)(a)}$$

```
• R2 := limit(rr, m = 0);
```

$$R2 := \frac{1 + 3\,\mathrm{D}(y)(a)^2 + 3\,\mathrm{D}(y)(a)^4 + \mathrm{D}(y)(a)^6}{D^{(2)}(y)(a)^2}$$

An **expand** on H and K will reproduce the results of $q27$ expanded, while factoring $R2$ and taking the square root will reproduce the standard formula for the radius of curvature.

- `expand(H);`

$$-\frac{\mathrm{D}(y)(a)}{D^{(2)}(y)(a)}+a-\frac{\mathrm{D}(y)(a)^3}{D^{(2)}(y)(a)}$$

- `expand(K);`

$$\frac{1}{D^{(2)}(y)(a)}+\mathrm{y}(a)+\frac{\mathrm{D}(y)(a)^2}{D^{(2)}(y)(a)}$$

- `simplify(sqrt(factor(R2)), symbolic);`

$$\frac{(\mathrm{D}(y)(a)^2+1)^{3/2}}{D^{(2)}(y)(a)}$$

3. Evolutes

The curve traced out by the moving center of Curvature is called the evolute. Before studying why this curve is interesting we'l construct the evolute for the curve $y(x) = x^2$. Begin by picking a curve and calculating its derivatives. From Section 2, we still have this curve assigned to f, so

- `fp := diff(f, x);`

$$fp := 2\,x$$

- `fpp := diff(f, x, x);`

$$fpp := 2$$

Define the parametric representation of (h, k) as determined in Section 2.2.

- `U := x - fp*(1 + fp^2)/fpp;`

$$U := x - x\,(1+4\,x^2)$$

- `V := f + (1 + fp^2)/fpp;`

$$V := 3\,x^2+\frac{1}{2}$$

Now, plot the parabola and its evolute. The scaling can also be done interactively after the plot is created.

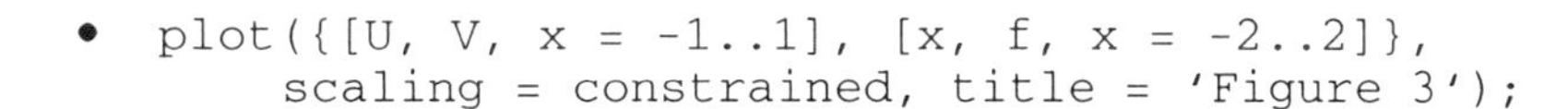

```
• plot({[U, V, x = -1..1], [x, f, x = -2..2]},
      scaling = constrained, title = 'Figure 3');
```

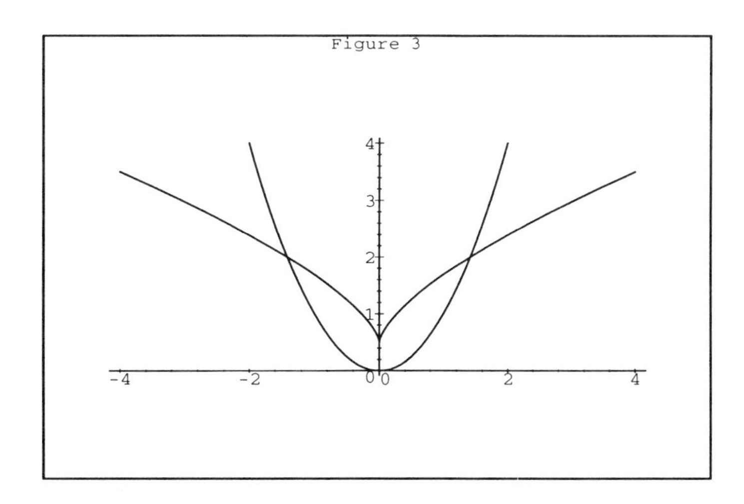

We continue this exploration by constructing three tangents to the evolute. We pick three values for the parameter x and store them in the list XX.

```
• XX := [1/2, 1, 3/2];
```

$$XX := \left[\frac{1}{2}, 1, \frac{3}{2}\right]$$

To compute the slopes of these tangent lines we will need to express the derivative of the evolute via the evolute s parametric representation.

```
• M := diff(V, x)/diff(U, x);
```

$$M := -\frac{1}{2}\frac{1}{x}$$

Computing the actual coordinates on the evolute and the slopes, and then constructing the tangent lines can all be done in a loop. Note how the colon on the loop terminator (**od**) suppresses output of all commands in the loop except for the **print** command.

```
• for j from 1 to 3 do
    x.j := subs(x = XX[j], U);
    y.j := subs(x = XX[j], V);
    m.j := subs(x = XX[j], M);
    t.j := y = y.j + m.j*(x - x.j);
    print(t.j);
  od:
```

$$y = \frac{3}{4} - x$$

$$y = \frac{3}{2} - \frac{1}{2}x$$

$$y = \frac{11}{4} - \frac{1}{3}x$$

Now, construct a graph with each of these three tangent lines, the evolute, and the parabola. Parametric representations give greatest control over the domains of the individual functions.

```
plot({[x, rhs(t1), x = -1..1], [x, rhs(t2), x = -5..1],
  [x, rhs(t3), x = -12..2], [U, V, x = 0..3/2],
  [x, f, x = 0..2]}, scaling=constrained, title = 'Figure 4');
```

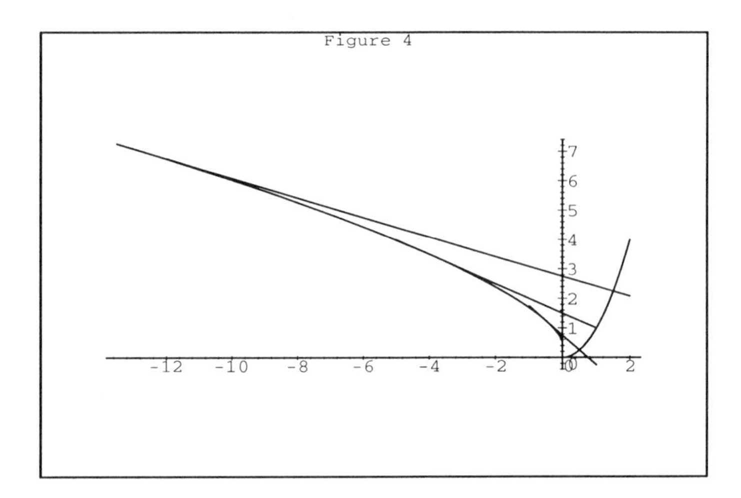

This construction seems to suggest that tangents on the evolute are normals on the parabola. Let's try a direct frontal attack on this hypothesis. For the curve given by $y = y(x)$, the parametric representation of the evolute is now known in terms of x, $y(x)$, $y'(x)$, and $y''(x)$. Thus, in Maple we write the first derivative

```
yp := diff(y(x), x);
```

$$yp := \frac{\partial}{\partial x}\, \mathrm{y}(x)$$

and the second derivative

```
ypp := diff(y(x), x, x);
```

$$ypp := \frac{\partial^2}{\partial x^2}\, \mathrm{y}(x)$$

Then, the parametric representation of the evolute is given by

```
he := x - yp*(1 + yp^2)/ypp;
```

$$he := x - \frac{\left(\frac{\partial}{\partial x}\, \mathrm{y}(x)\right)\left(1 + \left(\frac{\partial}{\partial x}\, \mathrm{y}(x)\right)^2\right)}{\frac{\partial^2}{\partial x^2}\, \mathrm{y}(x)}$$

```
ke := y(x) + (1 + yp^2)/ypp;
```

$$ke := \mathrm{y}(x) + \frac{1 + \left(\frac{\partial}{\partial x}\, \mathrm{y}(x)\right)^2}{\frac{\partial^2}{\partial x^2}\, \mathrm{y}(x)}$$

Tangents on the evolute require that we compute the derivative via the parametric representation.

- `q29 := diff(ke, x)/diff(he, x);`

$$q29 := \frac{3\left(\frac{\partial}{\partial x}\mathrm{y}(x)\right) - \frac{\left(1+\left(\frac{\partial}{\partial x}\mathrm{y}(x)\right)^2\right)\left(\frac{\partial^3}{\partial x^3}\mathrm{y}(x)\right)}{\%1^2}}{-3\left(\frac{\partial}{\partial x}\mathrm{y}(x)\right)^2 + \frac{\left(\frac{\partial}{\partial x}\mathrm{y}(x)\right)\left(1+\left(\frac{\partial}{\partial x}\mathrm{y}(x)\right)^2\right)\left(\frac{\partial^3}{\partial x^3}\mathrm{y}(x)\right)}{\%1^2}}$$

$$\%1 := \frac{\partial^2}{\partial x^2}\mathrm{y}(x)$$

If this simplifies to $-1/y'(x)$, then we have established that tangents on the evolute are normals on the original curve.

- `q30 := simplify(q29);`

$$q30 := -\frac{1}{\frac{\partial}{\partial x}\mathrm{y}(x)}$$

It seems that our hypothesis is true.

4. Parallel Curves

We have discovered that the evolute of a given curve $y(x)$ is the locus of the centers of curvature for the original curve $y(x)$. Each such center of curvature lies on a normal to $y(x)$, at the varying distance R, the radius of curvature, from the given curve. In fact, the parametric formulas for (h, k), the cartesian coordinates of this locus of centers, are contained in the expressions $q27expanded$. The independent variable x in $y(x)$ is the parameter in these expresions which can be written as

$$h = x - y'(x)\left(\frac{1+y'(x)^2}{y''(x)}\right)$$

and

$$k = y(x) + \frac{1+y'(x)^2}{y''(x)}$$

Noticing that the variable radius of curvature is $R = \frac{(1+y'(x)^2)^{3/2}}{y''(x)}$ we write the parametric equations for the evolute as

$$h = x - y'(x)\left(\frac{R}{\sqrt{1+y'(x)^2}}\right)$$

and

$$k = y(x) + \frac{R}{\sqrt{1 + y'(x)^2}}$$

Finally, we ask what happens if we take R as a constant d. Clearly, we should generate a curve parallel to $y(x)$ and at a distance d from it, the distance d being measured along the normal to $y(x)$.

We now implement an example in Maple. Since the letter f presently points to x^2, we begin by returning f to variable status.

```
• f := 'f';
```

$$f := f$$

Since we want to invoke the formulas for h and k with several values of the distance d, we set up the relevant expressions as functions of d and of a curve parameter, t.

```
• X := (t,d) -> t  - d*D(f)(t)/sqrt(1 + D(f)(t)^2);
```

$$X := (t, d) \to t - \frac{d\,\mathrm{D}(f)(t)}{\mathrm{sqrt}\,(1 + \mathrm{D}(f)(t)^2)}$$

```
• Y := (t,d) -> f(t) + d/sqrt(1 + D(f)(t)^2);
```

$$Y := (t, d) \to \mathrm{f}(t) + \frac{d}{\mathrm{sqrt}\,(1 + \mathrm{D}(f)(t)^2)}$$

Use of the notation $D(f)(t)$ for differentiation is predicated on f representing a function rather than an expression. Hence,

```
• f := x -> x^2;
```

$$f := x \to x^2$$

Having specified a function $f(x)$, we look at the consequences in the formulas for the parallel curves.

```
• X(t, d);
```

$$t - 2\frac{d\,t}{\sqrt{1 + 4\,t^2}}$$

and

```
• Y(t, d);
```

$$t^2 + \frac{d}{\sqrt{1 + 4\,t^2}}$$

We seem to have accomplished that part of the task successfully. Now, let's see if we can finish the example just as efficiently. We want to plot a graph showing several parallel curves and the function $f(x)$. Let's say we want the distances d to be $-2, -1, 1,$ and 2. (A negative value puts the parallel curve on the side of $f(x)$ opposite from the center of curvature.) We put these values into a list called dd.

```
• dd := [seq(j, j = -2..2)];
```

$$dd := [-2, -1, 0, 1, 2]$$

Our list of distances contains the value zero. The curve that is zero units from $f(x)$ is $f(x)$ itself.

Since the formulas for the parallel curves are given parametrically, we'll set up the five lists $p1, \ldots, p5$ needed for parametric plotting via the following loop.

```
• for j from 1 to 5 do
      p.j := [X(t, dd[j]), Y(t, dd[j]), t = -3..3];
  od:
```

The colon on the terminating **od** suppresses output from the whole loop. The plot itself can now be generated via

```
• plot({p.(1..5)}, title = 'Figure 5');
```

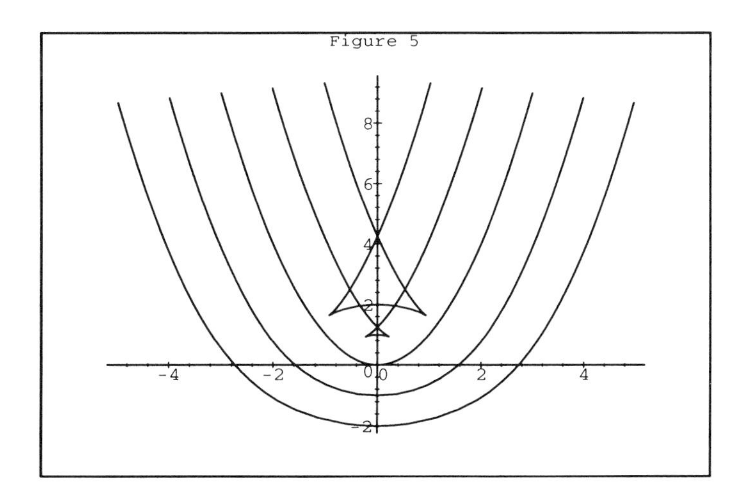

A final observation: each parallel curve is called an *involute* of the evolute. There is only one evolute, but there are an infinite number of involutes, each being a curve parallel to the original curve giving rise to the evolute. Thus, being an evolute is a property of the family of parallel curves which are in involution to the evolute. Since our excursion into the realm of the parallel curve was incidental, we leave it for the reader to work out a satisfactory demonstration that each member of a family of parallel curves generates the same evolute.

5. Envelopes

The excursion into the example of parallel curves is really an aside. The major issue is curvature, and that has led naturally to the evolute, from which the parallel curve follows as a "corollary."

We wish to explore the implications of Figure 4. The normals from the curve $y(x)$ are all tangent to the evolute. The evolute is therefore the envelope of the family of normals emanating from the curve $y(x)$. Can we recover the equations for the evolute by finding the envelope of the family of normals along $y(x)$?

First, however, we will have to make the capital investment in the concept and methodology of finding envelopes of a family of curves. The theory states that if $f(x, y, a) = 0$ represents a family of curves parametrized by a, then the equations $f = 0$ and $\frac{\partial f}{\partial a} = 0$ implicitly define the parametric equations $x = x(a)$, $y = y(a)$ for the envelope. In fact, the complete reasoning begins with writing the envelope as

$$x = X(a), \qquad y = Y(A) \tag{1}$$

and noting that the envelope has a common tangent with each member of the family

$$f(x, y, a) \;=\; 0 \tag{2}$$

Hence, for any common value of a, equations (1) will satisfy (2) so that (2), viewed as

$$f(X(a), Y(a), a) \;=\; 0 \tag{3}$$

leads, by differentiation with respect to a, to

$$f_X(X, Y, a)\frac{dX}{da} + f_X(X, Y, a)\frac{dY}{da} + f_a(X, Y, a) \tag{4}$$

where f_X, f_Y, and f_a are the partial derivatives of f with respect to X, Y, and a, respectively.

The slope of (1) at any point is

$$\frac{dy}{dx} \;=\; \frac{\frac{dY}{da}}{\frac{dX}{da}} \tag{5}$$

and the slope of a member of (2) at any point, using subscripts for partial differentiation, is

$$\frac{dy}{dx} \;=\; -\frac{f_x(x, y, a)}{f_y(x, y, a)} \tag{6}$$

The tangency of the envelope (1) and the members of the family (2) means the slopes in (5) and (6) are equal. Hence

$$\frac{\frac{dY}{da}}{\frac{dX}{da}} = -\frac{f_X(X(a), Y(a), a)}{f_Y(X(a), Y(a), a)} \tag{7}$$

or

$$f_X(X(a), Y(a), a)\frac{dX}{da} + f_Y(X(a), Y(a), a)\frac{dY}{da} = 0 \tag{8}$$

Finally, from (8) and (4) we deduce that $f_a(x, y, a) = 0$.

Curiosity leads us to implement this derivation in Maple. First, we ensure that the appropriate symbols are free variables.

- `f := 'f';`

$$f := f$$

- `X := 'X';`

$$X := X$$

- `Y := 'Y';`

$$Y := Y$$

Next, enter the equivalent of equation (2).

- `q31 := f(x, y, a);`

$$q31 := \mathrm{f}(x, y, a)$$

Then, implement the realization expressed in equation (3).

- `q32 := f(X(a), Y(a), a);`

$$q32 := \mathrm{f}(\mathrm{X}(a), \mathrm{Y}(a), a)$$

The differentiation leading to equation (4) is captured by

- `q33 := diff(q32 = 0, a);`

$$q33 := D_1(f)(\mathrm{X}(a), \mathrm{Y}(a), a)\left(\frac{\partial}{\partial a}\mathrm{X}(a)\right) + D_2(f)(\mathrm{X}(a), \mathrm{Y}(a), a)\left(\frac{\partial}{\partial a}\mathrm{Y}(a)\right) + D_3(f)(\mathrm{X}(a), \mathrm{Y}(a), a) = 0$$

The implicit differentiation expressed by equation (6) is implemented by the steps

- `q34 := f(x, y(x), a);`

$$q34 := \mathrm{f}(x, \mathrm{y}(x), a)$$

- `q35 := diff(q34 = 0, x);`

$$q35 := D_1(f)(x, \mathrm{y}(x), a) + D_2(f)(x, \mathrm{y}(x), a)\left(\frac{\partial}{\partial x}\mathrm{y}(x)\right) = 0$$

- `q36 := solve(q35, diff(y(x), x));`

$$q36 := -\frac{D_1(f)(x, \mathrm{y}(x), a)}{D_2(f)(x, \mathrm{y}(x), a)}$$

Equating the alternate views of the derivative on the envelope, contained in equation (7), is done via

- `q37 := diff(Y(a), a)/diff(X(a), a) = q36;`

$$q37 := \frac{\frac{\partial}{\partial a}\mathrm{Y}(a)}{\frac{\partial}{\partial a}\mathrm{X}(a)} = -\frac{D_1(f)(x, \mathrm{y}(x), a)}{D_2(f)(x, \mathrm{y}(x), a)}$$

However, as equation (7) was written, a rationalization of the arguments was implemented. Hence,

- `q38 := subs({x = X(a), y(x) = Y(a)}, q37);`

$$q38 := \frac{\frac{\partial}{\partial a}\mathrm{Y}(a)}{\frac{\partial}{\partial a}\mathrm{X}(a)} = -\frac{D_1(f)(\mathrm{X}(a), \mathrm{Y}(a), a)}{D_2(f)(\mathrm{X}(a), \mathrm{Y}(a), a)}$$

The manipulations that led from equation (7) to equation (8) are intuitive but Maple has to be explicitly told that (7) is to become (8).

- `q39 := numer(normal(lhs(q38) - rhs(q38))) = 0;`

$$q39 := D_1(f)(\mathrm{X}(a), \mathrm{Y}(a), a)\left(\frac{\partial}{\partial a}\mathrm{X}(a)\right) + D_2(f)(\mathrm{X}(a), \mathrm{Y}(a), a)\left(\frac{\partial}{\partial a}\mathrm{Y}(a)\right) = 0$$

Finally, the comparison between equations (8) and (4) can be implemented by simplifying, in Maple, $q33$ (the equivalent of equation (4)) subject to the side relation $q39$ (the equivalent of equation (8)).

- `simplify(q33, {q39});`

$$D_3(f)(\mathrm{X}(a), \mathrm{Y}(a), a) = 0$$

5.1 Example 1

Let's illustrate this enveloping technique by example. Remember, the reason for taking this detour is so that we can see if the equation of the evolute can be developed as the envelope of a family of normals.

The equation of a line at unit distance from the origin is, for any value of t, given by $x\cos(t) + y\sin(t) = 1$. Let's use Maple to create a picture before solving for the envelope analytically. Begin by entering the equation of this family of lines.

```
• q40 := x*cos(t) + y*sin(t) = 1;
```

$$q40 := x\cos(t) + y\sin(t) = 1$$

Next, solve for y.

```
• q41 := solve(q40, y);
```

$$q41 := -\frac{x\cos(t) - 1}{\sin(t)}$$

Now, create a function of the parameter t so that for each value of t we generate the equation of a distinct line. Since the expression forming the rule of the function already exists in Maple, we use the **unapply** command to convert this expression into a function.

```
• f := unapply(q41, t);
```

$$f := t \to -\frac{x\cos(t) - 1}{\sin(t)}$$

Since 2π is approximately 6, a list of 30 "uniformly" spaced values of t between 0 and 2π is given by

```
• T := [seq(j/5, j=1..30)]:
```

A plot of all these lines in a 2-by-2 window is created with

```
• plot({seq(f(T[j]), j = 1..30)}, x = -2..2, y = -2..2,
       scaling = constrained, title = 'Figure 6');
```

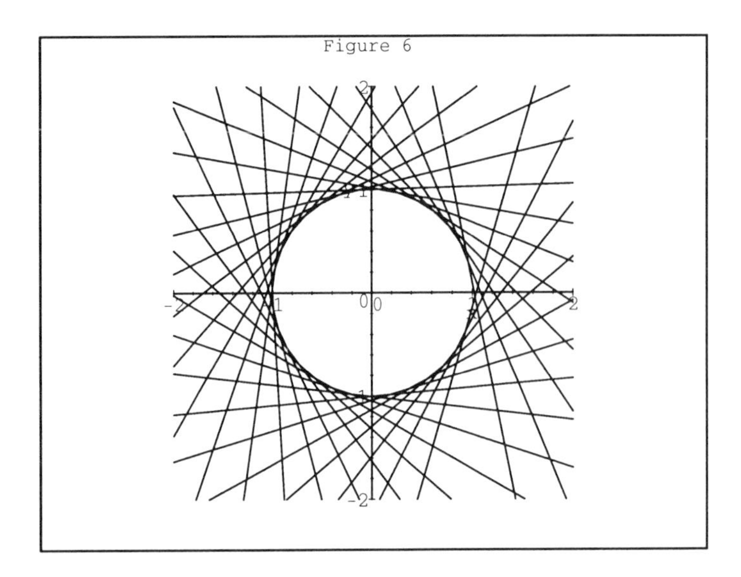

The analytic result we seek is carried out in Maple via

```
• q42 := diff(q40,t);
```

$$q42 := -x \sin(t) + y \cos(t) = 0$$

The equations $f(x, y, t) = 0$ and $f_t(x, y, t) = 0$ define the envelope parametrically via $x = x(t)$ and $y = y(t)$. Hence,

```
• q43 := solve({q40, q42}, {x, y});
```

$$q43 := \left\{ x = \frac{\cos(t)}{\cos(t)^2 + \sin(t)^2}, y = \frac{\sin(t)}{\cos(t)^2 + \sin(t)^2} \right\}$$

Clearly, this simplifies.

```
• simplify(q43);
```

$$\{ y = \sin(t), x = \cos(t) \}$$

Since this is easily recognized as the parametric representation of a circle, we move on to a new example.

5.2 Example 2

Let's find the envelope of all lines of unit length with endpoints on the coordinate axes. Figure 7 below is used in a derivation of the governing equation of this family of lines. It is created in Maple with the following commands.

```
• OA := [[0,0],[5,0]]:
```

```
OB := [[0,0],[0,6]]:
AB := [[4,0],[0,5]]:
OC := [[0,0],[100/41,80/41]]:
CF := [[100/41,0],[100/41,80/41]]:
DF := [[100/41,0],[2500/1681,2000/1681]]:
p3 := plot({OA,OB,AB,OC,CF,DF},
     scaling = constrained, axes = none, title = 'Figure 7'):
p4 := textplot({[0,-.2,'O'],[4,-.2,'A'],[-.2,5,'B'],
       [3.5,2.3,'C:(x,y)'],[1.4,1.5,'D'],
       [100/41,-.2,'F'],[2,3,'d'],[.2,4.5,'t'],
       [.7,.35,'t']}):
display([p3,p4]);
```

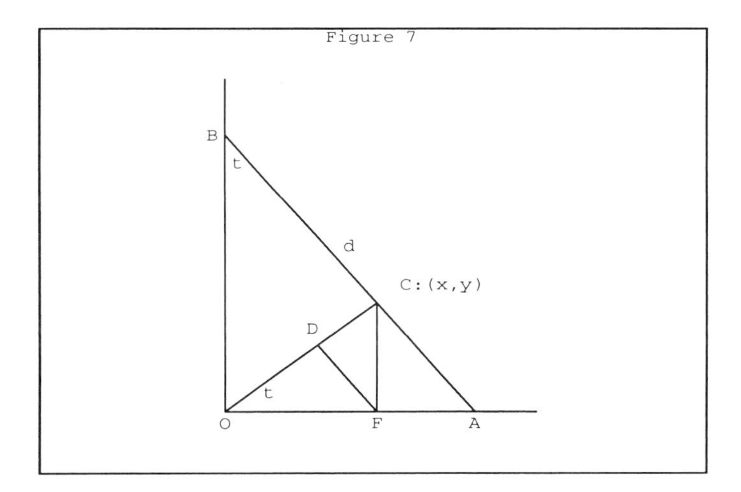

Let the endpoints of the line, here given the arbitrary length d , be A and B . Draw the perpendicular OC from the origin O to the line. The cartesian coordinates of point C are taken as (x, y). Drop a perpendicular from C to the x-axis at F. Drop a perpendicular FD from F to the segment OC.

Distance OF is x, and distance FC is y; hence, $OD = x\cos(t)$ and $DC = y\sin(t)$, so $OC = x\cos(t) + y\sin(t)$. But from triangle OAB we have that OB is $d\cos(t)$, so in triangle OBC, length OC is $OB\sin(t)$ or $d\cos(t)\sin(t)$. Hence, the equation we seek is $x\cos(t) + y\sin(t) - d\cos(t)\sin(t) = 0$.

Take $d = 1$ and enter this equation into Maple in an effort to repeat the calculations of the previous example.

- `q44 := x*cos(t) + y*sin(t) - cos(t)*sin(t);`

$$q44 := x\cos(t) + y\sin(t) - \cos(t)\sin(t)$$

As before, isolate y.

- `q45 := solve(q44, y);`

$$q45 := -\frac{x\cos(t) - \cos(t)\sin(t)}{\sin(t)}$$

Convert this expression into a function of the parameter t so that we can again plot representative members of the family of lines.

```
• f := unapply(q45, t);
```

$$f := t \to -\frac{x\cos(t) - \cos(t)\sin(t)}{\sin(t)}$$

Now, create a list of t-values at which lines in the family will be drawn.

```
• T := [seq(Pi/2*j/10, j = 1..10)]:
```

Plot these lines in a 1-by-1 window.

```
• plot({seq(f(T[j]), j = 1..10)}, x = 0..1, y = 0..1,
       title = 'Figure 8');
```

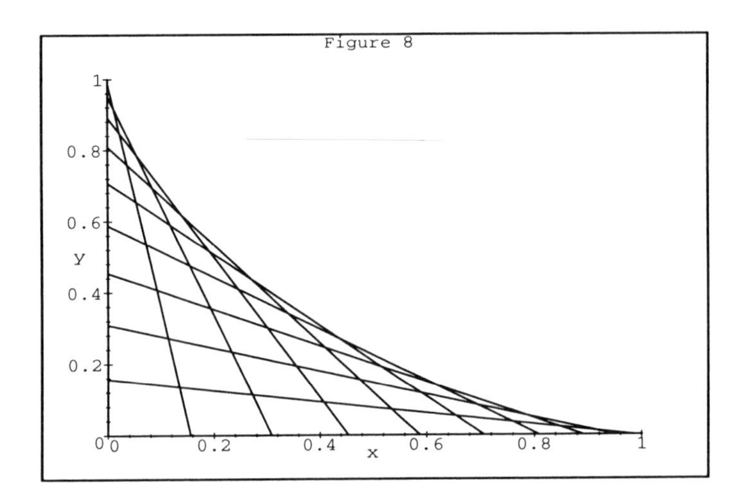

We continue with the analytical extraction of the equations of the envelope. Using Maple we obtain the analog of $f_t(x, y, t) = 0$ via

```
• q46 := diff(q44, t);
```

$$q46 := -x\sin(t) + y\cos(t) + \sin(t)^2 - \cos(t)^2$$

and solve the pair of equations $f(x, y, t) = 0$ and $f_t(x, y, t) = 0$.

```
• q47 := solve({q44, q46}, {x, y});
```

$$q47 := \left\{y = \frac{\cos(t)^3}{\cos(t)^2 + \sin(t)^2}, x = \frac{\sin(t)^3}{\cos(t)^2 + \sin(t)^2}\right\}$$

It s clear what x and y are, but having Maple admit it requires

```
• q48 := subs(sin(t)^2 + cos(t)^2 = 1, q47);
```

$$q48 := \left\{y = \cos(t)^3, x = \sin(t)^3\right\}$$

It is interesting that the envelope is a hypocycloid. In fact, if these parametric results are plotted for t in $[0, 2]$, we obtain the full graph of the hypocycloid shown in Figure 9 below. In Maple we extract the parametric equations from $q48$ via

- `xt := subs(q48, x);`

$$xt := \sin(t)^3$$

- `yt := subs(q48, y);`

$$yt := \cos(t)^3$$

whence

- ```
  plot([xt, yt, t = 0..2*Pi], scaling = constrained,
        title = 'Figure 9');
  ```

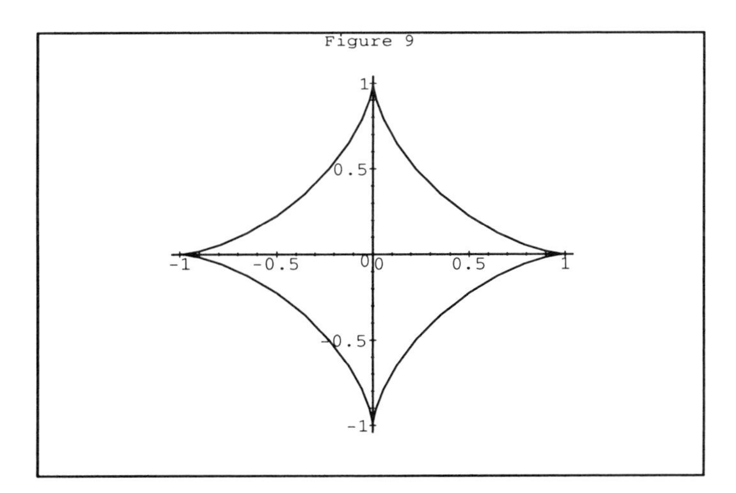

5.3 Evolute As Envelope

Finally, let's see if we can obtain the equation of an evolute by finding the envelope of the family of normals on the given curve $y(x)$. We'll use Maple to do the algebra. In addition, we'll make use of of the **alias** facility to compact some of the results.

First we make sure the symbols we want to use are available.

- `f := 'f';`

$$f := f$$

- `p := 'p';`

$$p := p$$

Maple's **alias** command allows us to establish suitable abbreviations. The response to an **alias** command is the list of all aliased items. To remove an **alias** y, you use the syntax **alias(y=y)**.
  ```

We begin by pointing f , p, and P to $y(x)$, $y'(x)$ and $y''(x)$ respectively.

- `alias(f = y(x));`

$$I, f$$

- `alias(p = diff(y(x), x));`

$$I, f, p$$

- `alias(P = diff(y(x), x, x));`

$$I, f, p, P$$

The family of normals is given by

- `q49 := Y - f = -1/p*(X - x);`

$$q49 := Y - f = -\frac{X - x}{p}$$

Since the parameter in this family is x , the step corresponding to forming the equation $f_a(X, Y, a) = 0$ is

- `q50 := diff(q49, x);`

$$q50 := -p = \frac{(X - x)\,P}{p^2} + \frac{1}{p}$$

Solving the equations analogous to $f(x, y, a) = 0$ and $f_a(X, Y, a) = 0$ for $x = X(a)$ and $y = Y(a)$ is accomplished by

- `q51 := solve({q49, q50}, {X, Y});`

$$q51 := \left\{X = \frac{-p^3 + x\,P - p}{P}, Y = \frac{p^2 + 1 + P\,f}{P}\right\}$$

A modest rearrangement is possible via

- `q52 := collect(q51, P);`

$$q52 := \left\{Y = f + \frac{1 + p^2}{P}, X = x + \frac{-p^3 - p}{P}\right\}$$

Careful inspection of $q52$ shows that we have recovered the parametric representation, given by the equations

$$h = x - y'(x)\frac{1 + y'(x)^2}{y''(x)}$$

and

$$k = y(x) + \frac{1 + y'(x)^2}{y''(x)}$$

for the evolute of the curve $y = y(x)$.

Once again we remind the reader that this excursion into the evolute was taken as an illustration of the enveloping process. In particular, we have derived the equation of the evolute as the envelope of normals on the curve $y = y(x)$. Our interest in the evolute is as the locus of the center of curvature for the curve defined by $y = y(x)$. The fundamental investigation responsible for these activities was that of discovering a characterization of the circle of curvature. Our path to that insight led us to the center of curvature; the locus of the center of curvature is the evolute.

In the final section of this unit we apply the idea of enveloping to a problem posed in a popular mathematics journal. We've answered our original question about the circle of curvature, but this last study will tie in with the examples met while examining the technique of enveloping.

6. The Bi-Fold Door Problem

The final Maple activity in this study of curvature, and its consequence is a solution, by enveloping, of the Bi-Fold Door Problem recently posed in the *Mathematics Magazine* (Vol. 66, No. 3, Problem #1427, June 1993). The problem has a bi-fold door of total (closed) width $2a$ dragging over a rug in such a way as to cause wear. The problem is to determine the shape of the region of wear as the door is repeatedly opened and closed.

We use Figure 10 below, created by the given Maple code, to establish our frame of reference.

```
• p5 := plot({[[0,0],[5,0]],[[0,0],[0,3]],[[2,0],[2,2]],[[0,0],
            [2,2]], [[2,2],[4,0]]},
            scaling = constrained, axes = none,
            title = 'Figure 10'):
  p6 := textplot({[0,-.2,'O'],[4,-.2,'A'],[2,2.2,'B'],
                 [1,1.3,'a'],[3,1.3,'a'],
                 [.35,.2,'t']}):
  display([p5,p6]);
```

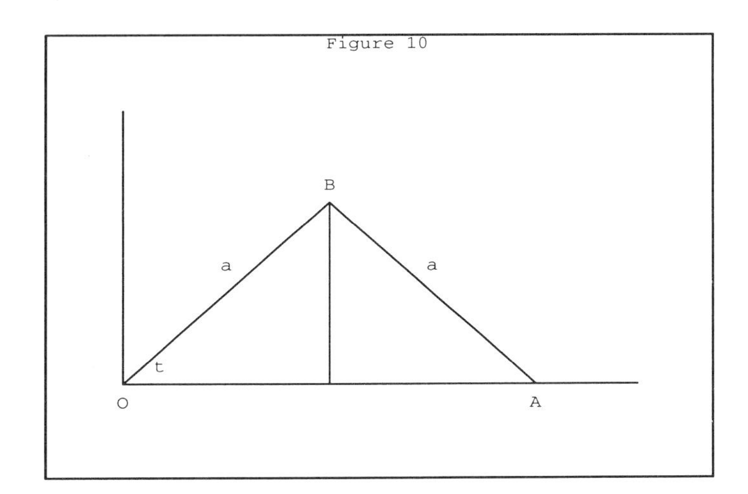

The main hinge for the door is at O, and the secondary hinge is at B. Point A moves horizontally in the glide OA. Angle t , t in $[0, \pi/2]$, measures the amount by which the door has been opened. Coordinates for point B are $(a\cos(t), a\ sin(t))$ while coordinates for point A are $(2a\cos(t), 0)$.

It is possible to devise a graphical "solution" to this problem via Maple. We implement that thinking by defining a function $z(t)$ which gives, at each t in $[0, \pi/2]$, the data structure Maple needs for a plot of a snapshot of the moving door. Thus, assuming a to be 1,

```
• z := t -> [ [0,0], [cos(t), sin(t)], [2*cos(t), 0] ];
```

$$z := t \to [[0,0],[\cos(t),\sin(t)],[2\cos(t),0]]$$

We can decide on a sequence of times t at which the snapshots are to be taken. These times, stored in a list called T , are taken as uniformly distributed in the interval $[0, \pi/2]$.

```
• T := [seq(Pi/2*j/15,j=0..15)]:
```

A plot of all 16 snapshots gives a strong visual representation of the area of wear beneath the dragging bi-fold door.

```
• plot({seq(z(t), t = T)}, scaling = constrained,
        title = 'Figure 11');
```

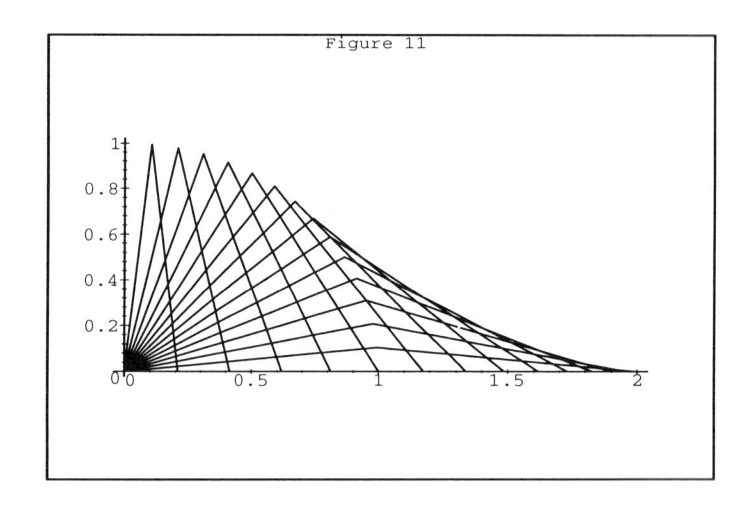

A more interesting graphical insight can be devised if we animate the snapshots just displayed. To assist with this activity we define the function $zz(n)$, which will create the graph of the nth snapshot in Figure 11.

```
• zz := n -> plot(z(T[n]));
```

$$zz := n \to \text{plot}\ (\text{z}\ (T_n))$$

The command for generating the animation is given below. Paper is an inappropriate medium for showing an animation, so the continuation of this line of investigation is left to the reader. Note that the **display** command used is found in the plots package.

```
• display([seq(zz(n), n = 1..16)], insequence = true);
```

6.1 Analytic Solution

We return to the task of finding an analytic solution to the Bi-Fold Door Problem. Looking at Figure 10 and thinking of door segment AB as generating a family of lines, we seek a representation of this collection of segments. In fact, by using Maple's geometry package we can implement the necessary coordinate geometry. Since the geometry package has the quirk that the variables x and y are not available to the user, be sure that neither x nor y have been assigned values before loading the geometry package.

- `with(geometry):`

Within the geometry package give the name "hinge" to the Maple data structure "point," whose coordinates are those of the secondary hinge at B.

- `point(hinge, [a*cos(t), a*sin(t)]);`

$$hinge$$

The response is the name of the named point. Actually, this is the name of a table which contains the information about this object, and the actual coordinates can be recovered from this table if needed. This recovery will be illustrated momentarily.

Enter the name "doorglide" as the name of point A.

- `point(doorglide, [2*a*cos(t), 0]);`

$$doorglide$$

Next, have Maple generate the equation of the line connecting points A and B. The name of the table holding the information about this line segment is $q53$.

- `line(q53, [hinge, doorglide]);`

$$q53$$

To see the equation of the segment AB, we actually use the syntax of a Maple table. This equation is of the form $f(x, y, t) = 0$.

- `q54 := q53[equation];`

$$q54 := a\sin(t)\,x + a\cos(t)\,y - 2\,a^2\sin(t)\cos(t) = 0$$

The equivalent of the equation $f_t(x, y, t) = 0$ is

- `q55 := diff(q54, t);`

$$q55 := a\cos(t)\,x - a\sin(t)\,y - 2\,a^2\cos(t)^2 + 2\,a^2\sin(t)^2 = 0$$

We next need to solve the equivalents of the equations $f(x, y, t) = 0$ and $f_t(x, y, t) = 0$ for $x = x(t)$ and $y = y(t)$.

```
• q56 := solve({q54, q55}, {x, y});
```

$$q56 := \left\{ y = 2\,\frac{a\sin(t)^3}{\cos(t)^2+\sin(t)^2},\, x = 2\,\frac{a\cos(t)^3}{\cos(t)^2+\sin(t)^2} \right\}$$

The attentive reader will recognize this result. It appeared earlier in the discussion of the hypocycloid where we applied the exact same simplification as we will now.

```
• q57 := subs(sin(t)^2 + cos(t)^2 = 1, q56);
```

$$q57 := \left\{ y = 2\,a\sin(t)^3,\, x = 2\,a\cos(t)^3 \right\}$$

Extracting these parametric equations from the set $q57$ we get

```
• xt := subs(q57, x);
```

$$xt := 2\,a\cos(t)^3$$

```
• yt := subs(q57, y);
```

$$yt := 2\,a\sin(t)^3$$

It is tempting to plot the hypocycloid represented parametrically by xt and yt. However, any naivety here will lead to paradoxical results. Suppose, for example, we set $a = 1$ and plot for t in $[0, \pi/2]$.

```
• a := 1;
```

$$a := 1$$

```
• plot([xt, yt, t = 0..Pi/2], scaling = constrained,
       title = 'Figure 12');
```

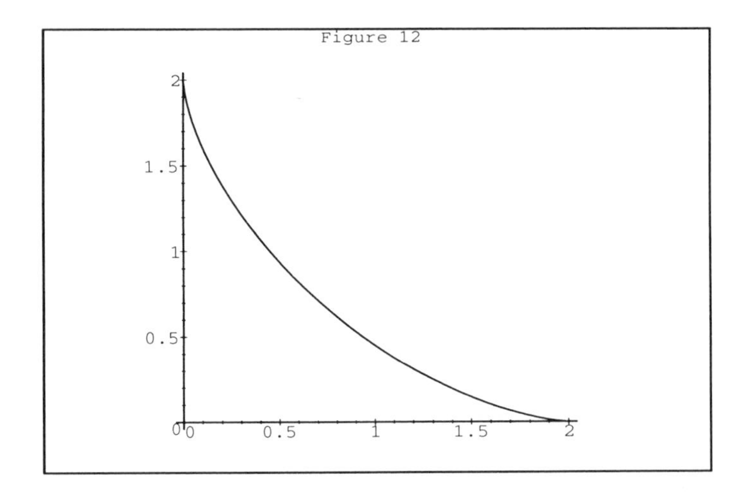

Figure 12 is patently wrong, for it shows wear where there is never passage of the door. To make a long story short, we have a domain problem. The domain for the equations of the hypocycloid is not t in $[0, \pi/2]$. Our domain is smaller because the point of contact of the family of lines and its envelope must remain attached to the right-hand segment of the door, that is, below the level of the hinge at B. Consequently, there is an inherent constraint on yt (it can not exceed $a\sin(t)$) that was never employed. We begin by looking at the graphs of yt and $a\sin(t)$.

```
• plot({yt, a*sin(t)}, t = 0..Pi/2, title = 'Figure 13');
```

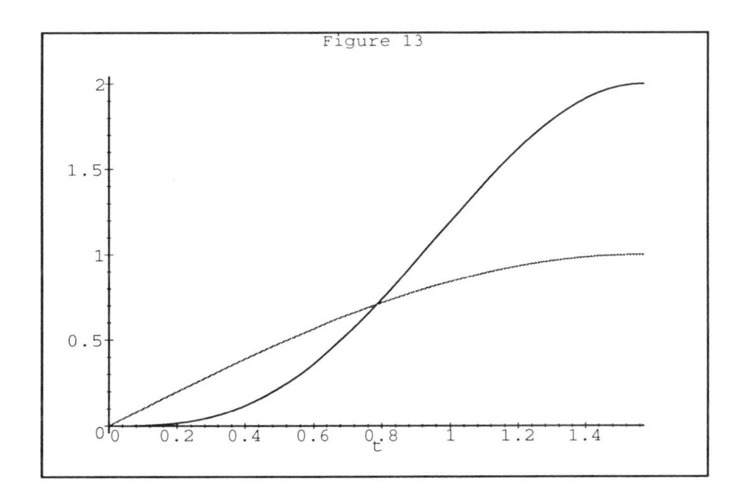

Figure 13 suggests that we solve the equation $yt = a\sin(t)$ to find the critical value of t at which the constraining inequality no longer holds. We first restore a to its variable status.

```
• a := 'a';
```

$$a := a$$

```
• solve(yt = a*sin(t), t);
```

$$0, \frac{1}{4}\pi, -\frac{1}{4}\pi$$

Since it is obvious that the critical value of t is $\pi/4$ we will simply join together an arc of a circle for which t is in $[\pi/4, \pi/2]$ and that segment of the hypocycloid for which t is in $[0, \pi/4]$. This composite curve will bound the region of wear beneath the dragging bi-fold door.

Plotting requires we set $a = 1$, say.

```
• a := 1;
```

$$a := 1$$

Piecewise defined functions are best graphed parametrically.

```
• plot({[xt, yt, t = 0..Pi/4],
      [a*cos(t), a*sin(t), t = Pi/4..Pi/2]},
        scaling = constrained, title = 'Figure 14');
```

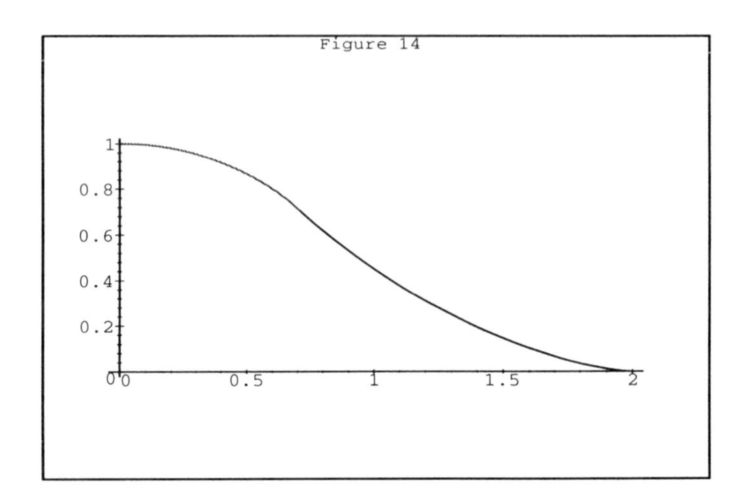

6.2 A Curiosity Resolved

We end with one final question: why did the hypocycloid appear in the Bi-Fold Door Problem? We initially obtained the hypocycloid as the solution to a geometric problem of enveloping the family of lines of unit length whose ends were along the coordinate axes. The bi-fold door yielded the hypocycloid via a family of line segments, only one end of which touched the x-axis. What is the connection?

We resolve this question by referring to Figure 15, an extension of Figure 10 that is created by the following Maple code.

```
• p7 := plot({[[0,0],[3,0]],[[0,0],[0,3]],[[2,0],
              [0,2]],[[0,0],[1,1]]},
              scaling = constrained, axes = none,
              title = 'Figure 15'):
  p8 := textplot({[0,-.2,'O'],[2,.2,'A'],[.2,2,'C'],
                 [1.1,1.2,'B'],[.2,1.7,'z'],[.2,.4,'w'],
                 [.8,1.1,'y'], [1,.8,'x'], [.35,.2,'t'],
                 [1.7,.2,'t']}):
```

```
• display([p7,p8]);
```

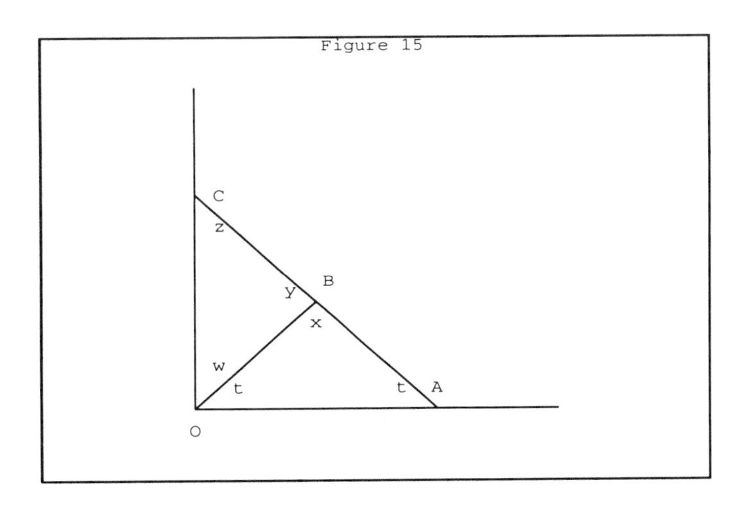

In Figure 15, we extend the segment AB until it hits the y-axis in point C. Angles AOB and OAB are both labeled t since triangle OAB is isosceles. Angles x and y are supplementary angles, while angles t and w and angles t and z are complimentary pairs. In fact, the connections between angles t, w, z, and y are easily expressed algebraically in Maple via

- `e1 := t + w = Pi/2;`

$$e1 := t + w = \frac{1}{2}\pi$$

- `e2 := t + z = Pi/2;`

$$e2 := t + z = \frac{1}{2}\pi$$

- `e3 := y = 2*t;`

$$e3 := y = 2\,t$$

- `e4 := y = Pi - z - w;`

$$e4 := y = \pi - z - w$$

The implications of these four relationships are then extracted via

- `solve({e.(1..4)},{t,w,z,y});`

$$\left\{z = w, y = \pi - 2\,w, t = \frac{1}{2}\pi - w, w = w\right\}$$

This solution reveals the important fact that $w = z$. Thus, angles BOC and OCB are equal, making triangle OBC isosceles. Hence, $AB = OB = BC$. The right half of the bi-fold door behaves as half of the line segment that generated the hypocycloid in the earlier example. That is why the Bi-Fold Door Problem generates the same solution: the hypocycloid. Interestingly enough, the Bi-Fold Door Problem also makes a statement about the hypocycloid:

hinging a rod of length a between the origin and the midpoint of a rod of length 2 a allows the longer rod to generate, by enveloping, the hypocycloid.

We started with curvature and angled our way to an interesting curve!

Unit 26: The Lagrange Multiplier, Part One

Technology allows us to examine, in a way not posible with pencil and paper, the ideas behind the Lagrange Multiplier technique. The following problem is a standard textbook exercise involving, if we will, Lagrange Multipliers. The solution will be embellished with additional explorations into the meaning of the calculations.

Find the extreme values of the function

$$f(x, y) = x^2 + y^2$$

on the line $x + 2y = 5$.

Geometrically, $f(x, y)$ represents a surface and the constraint is a line. Perhaps if we looked at a contour plot of $f(x, y)$ and a graph of the constraining line we might gain some insight into the structure of this problem.

First, a contour plot of $f(x, y)$. We'll use the **contourplot** command in the plots package. Begin by entering the function $f(x, y)$ and g, the constraint line.

- ```
  f := x^2 + y^2;
  ```

$$f := x^2 + y^2$$

- ```
  g := x + 2*y = 5;
  ```

$$g := x + 2\,y = 5$$

Then, load the plots package to gain access to the **contourplot** command.

- ```
  with(plots):
  contourplot(f, x = -5..5, y = -5..5,);
  ```

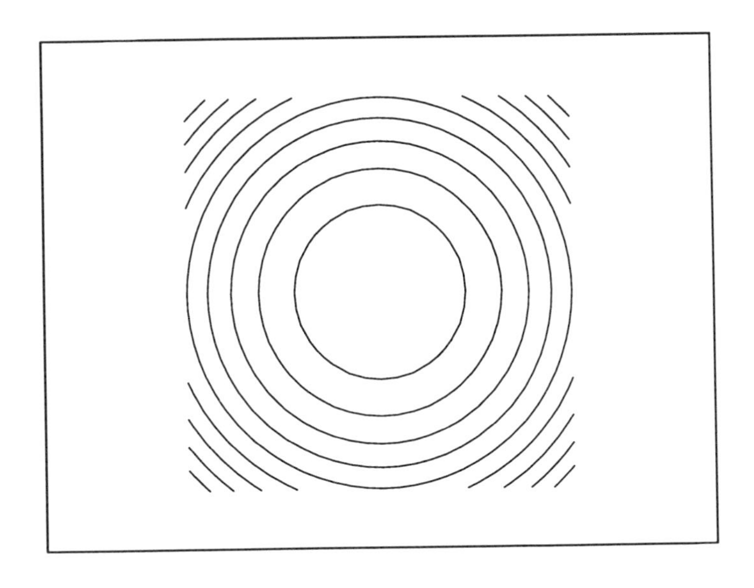
  ```

The interactive plot options were used to put the box around the figure and to cause both axes to have equal scaling. Alternatively, these features can be included from the keyboard as plot options via the usage

```
• contourplot (f, x = -5..5, y = -5..5, axes = boxed,
                    scaling = constrained);
```

In either event, this figure does not clearly indicate the values attained by $f(x, y)$ on the individual contours. Besides, this is a plot3d data-structure, and it will be incompatible with the 2d plot of the constraint $g(x, y) = 0$ that we are shortly going to obtain. Hence, we will create an alternate version of the contour map of the objective function $f(x, y)$.

In this alternative method we will generate individual implicit plots of $f(x, y) = constant$ and "paste" them into one graph. We'll use a loop to generate a series of plot data-structures. The colon after the loop-ending **od** is crucial. That colon suppresses the output of the whole loop. If it were not there, all the contents of the plot data-structures being generated would print to the screen.

```
• for k from 1 to 6 do
    f.k := implicitplot(f = k, x = -6..6, y = -6..6,):

  od:
```

Now paste all these individual contours into one plot. Note that again we assign the plot data-structure a name and therefore end the command with a colon to suppress the output of the plot data-structure itself.

```
• p1 := display([f.(1..6)]):
  p1;
```

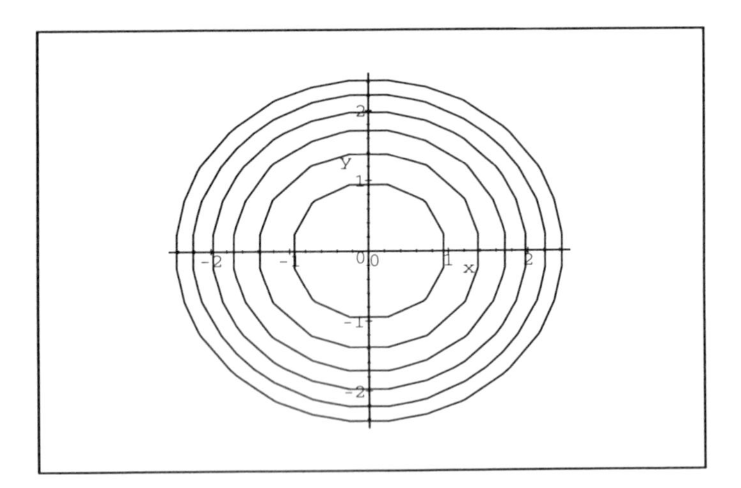

The axes can be "turned off" via the options window. In either event, it is against this contour map that we will display the plot of the constraint line.

```
• p2 := implicitplot(g, x = -3..3, y = -3..3):
  display([p1, p2]);
```

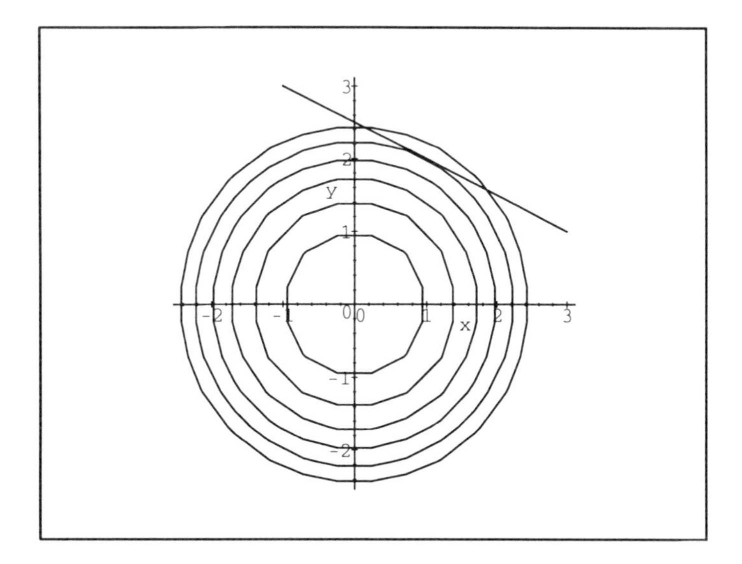

As you "walk along" the constraint line you "cut" level curves of $f(x, y)$. At any point on the constraint where "movement" will mean cutting a level curve, that point cannot be an extremum. At an extremum the values of the objective function $f(x, y)$ should be stationary. Hence, the extreme value must occur at a point where the constraint is tangent to a level curve, because at a point of tangency the constraint will not "cut through" a level curve.

Let's look more closely at this contention that at the extremum the constraint is tangent to a level curve.

- `display([f5, p2]);`

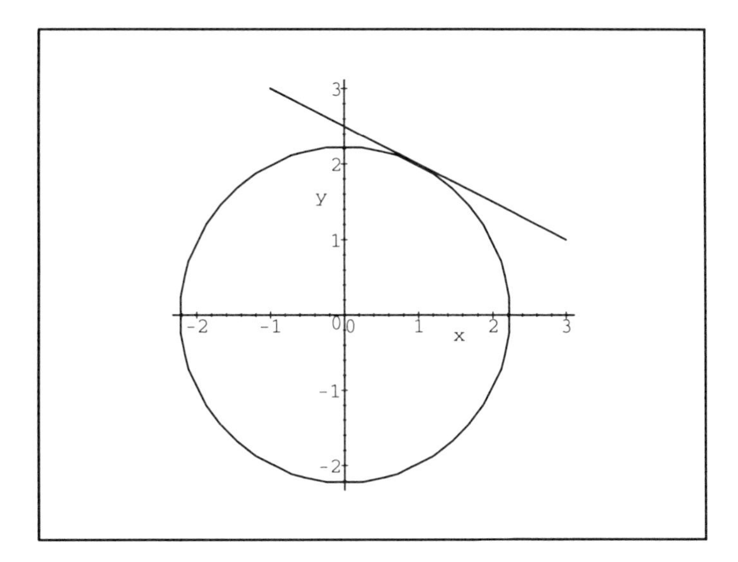

The figure does not rule out tangency, but the only sure verification is computational. We'll obtain the extreme value by eliminating one variable, a technique used in Calculus I. Thus, solve equation g for $x = x(y)$ and replace x in $f(x, y)$ with this expression for $x(y)$.

- `X := solve(g, x);`

$$X := -2\,y + 5$$

- `ff := subs(x = X, f);`

$$ff := (\,-2\,y + 5\,)^2 + y^2$$

Let's get a plot of $ff(y) = f(y(x), x)$.

- `plot(ff, y = -1..5);`

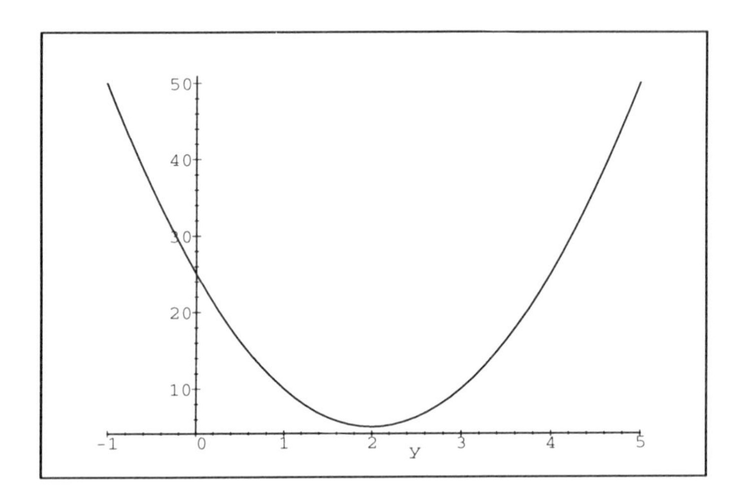

This plot suggests the extreme value is a minimum that occurs when $y = 2$.

The usual techniques of calculus lead to the exact location of this extremum.

- `q := diff(ff, y);`

$$q := 10\,y - 20$$

- `yc := solve(q, y);`

$$yc := 2$$

- `xc := subs(y = yc, X);`

$$xc := 1$$

Hence, we have found the extremum $(xc, yc) = (1, 2)$. Note that this point lies on the level curve $f(x, y) = 5 = f(1, 2)$. In fact,

- `subs(x = xc, y = yc, f);`

$$5$$

By inspection, the slope of the constraint line is $-1/2$. There is, however, a **slope** function in the student package. Rather than load the whole package, we can access and apply this one command as follows.

```
• student[slope](g, y(x));
```

$$\frac{-1}{2}$$

Next, we want, at the point $(1, 2)$, the slope on the level curve $f(x, y) = 5$. One route to this is implicit differentiation.

```
• q := subs(y = y(x), f);
```

$$q := x^2 + \mathrm{y}(x)^2$$

```
• q1 := diff(q, x);
```

$$q1 := 2\,x + 2\,\mathrm{y}(x)\left(\frac{\partial}{\partial x}\mathrm{y}(x)\right)$$

```
• q2 := solve(q1, diff(y(x), x));
```

$$q2 := -\frac{x}{\mathrm{y}(x)}$$

```
• q3 := subs({x = 1, y(x) = 2}, q2);
```

$$q3 := \frac{-1}{2}$$

That the slopes on the level curve and on the constraint line are the same indicates these two curves are tangent at the extremum. This recognition leads us to ask if there is some way of finding directly such a point of tangency. If so, we would have a new method for finding constrained extrema.

At a point of tangency between the constraint and the level curve, the gradient vectors have to be collinear. Let's check this for the example in question.

Load the linear algebra package for access to Maple's vector calculus functionality.

```
• with(linalg):

Warning: new definition for   norm
Warning: new definition for   trace
```

Compute gradient vectors for $f(x, y)$ and for the constraint.

```
• vf := grad(f, [x, y]);
```

$$vf := [\,2\,x \;\; 2\,y\,]$$

```
• vf1 := subs(x = 1, y = 2, op(vf));
```

$$vf1 := [\,2 \;\; 4\,]$$

Notice that the substitutions are made into the "operands" of the vector vf.

The constraint g was entered as an equation. Hence it is not a proper argument in the gradient command. In fact,

```
• grad(g, [x, y]);
```

Error, (in grad) wrong number (or type) of arguments

We could write

```
• G := lhs(g) - rhs(g);
```

$$G := x + 2\,y - 5$$

Hence,

```
• vg := grad(G, [x, y]);
```

$$vg := [\,1 \;\; 2\,]$$

By inspection we can see that $vf1 = 2vg$.

Let's look for the point of tangency by directly seeking a point at which the gradients of f and G are proportional. Matrix/ vector arithmetic is accomplished inside the **evalm** command.

```
• Q := evalm(grad(f, [x, y]) - m*grad(G, [x, y]));
```

$$Q := [\,2\,x - m \;\; 2\,y - 2\,m\,]$$

This is a pair of "equations" (the righthand side is assumed to be zero) in the unknowns x, y, and m. For a third equation we use the fact that x and y must lie on the constraint $G = 0$. All that remains is to extract the two "equations" contained in Q.

```
• Q[1];
```

$$2\,x - m$$

```
• Q[2];
```

$$2\,y - 2\,m$$

Hence, we solve the set of equations $\{Q[1] = 0, Q[2] = 0, G = 0\}$ for $\{x, y, m\}$ via

```
• solve({Q[1], Q[2], G}, {x, y, m});
```

$$\{\,x = 1, y = 2, m = 2\,\}$$

The typical treatment of the Lagrange Multiplier technique defines a new objective function $F(x, y, m)$ as $F = f - mG$. Then, partials with respect to x , y , and m generate the same three equations we solved above. (Differentiation of F with respect to m merely recovers the constraint equation G = 0.)

```
• F := f - m*G;
```

$$F := x^2 + y^2 - m\,(\,x + 2\,y - 5\,)$$

We can now apply the gradient operator to F, or we can differentiate F with respect to each of the variables x, y, and m.

- `QQ := grad(F, [x, y, m]);`

$$QQ := [\,2\,x - m \quad 2\,y - 2\,m \quad -x - 2\,y + 5\,]$$

- `solve({QQ[1], QQ[2], QQ[3]}, {x, y, m});`

$$\{\,x = 1, y = 2, m = 2\,\}$$

The set of equations can be also extracted from the vector QQ by the syntax

- `seq(QQ[k], k = 1..3);`

$$2\,x - m, 2\,y - 2\,m, -x - 2\,y + 5$$

Finally, we can clearly create these three equations individually via the syntax

- `e1 := diff(F, x);`

$$e1 := 2\,x - m$$

- `e2 := diff(F, y);`

$$e2 := 2\,y - 2\,m$$

- `e3 := diff(F, m);`

$$e3 := -x - 2\,y + 5$$

What remains is to apply these ideas to a more difficult problem. We do this in Units 27 and 28.

Unit 27: The Lagrange Multiplier, Part Two

This is a more computationally challenging example of the Lagrange Multiplier technique.

Find the extreme values of the function $f(x, y)$ on the curve defined implicitly by the constraint $g(x, y) = 0$ if

$$f(x, y) = x^2 + y^2$$

and

$$g(x, y) = 17y^2 + 12xy + 8x^2 - 46y - 28x + 17$$

Begin by entering the functions $f(x, y)$ and $g(x, y)$.

- ```
  f := x^2 + y^2;
  ```
  $$f := x^2 + y^2$$

- ```
  g := 17*y^2 + 12*x*y + 8*x^2 - 46*y - 28*x + 17;
  ```
  $$g := 17\,y^2 + 12\,x\,y + 8\,x^2 - 46\,y - 28\,x + 17$$

Let's get a contour plot of $f(x, y)$.

- ```
  with(plots):
  contourplot(f, x = -5..5, y = -5..5);
  ```

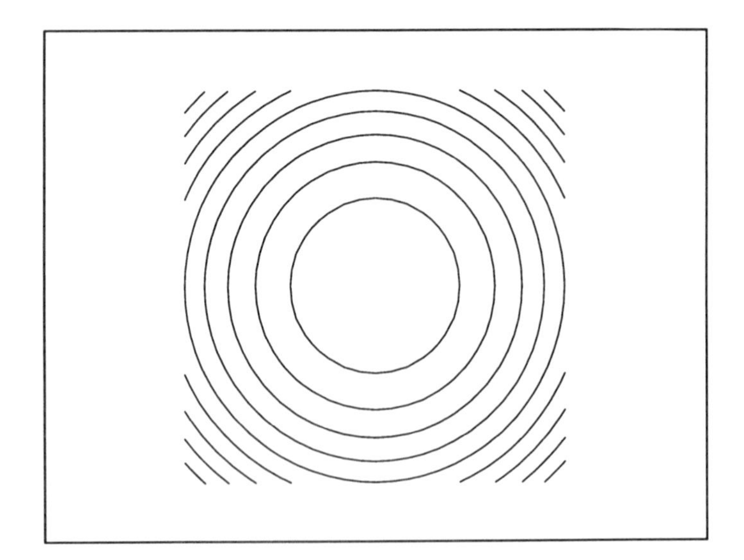

The interactive plot options were used to put the box around the figure and to cause both axes to have equal spacing. However, this figure does not indicate the values attained by $f(x, y)$ on the individual contours. Besides, this is a plot3d data structure, and it will be
  ```

incompatible with the implicit plot of the constraint $g(x, y) = 0$ that we are going to obtain shortly. Hence, we will create an alternate version of the contour map of the objective function $f(x, y)$.

In this alternative method we will generate individual implicit plots of $f(x, y) = constant$ and "paste" them into one graph. First, we will create R , a list of radii we want the circles $f(x, y) = constant$ to have.

```
• R:=[1/2,1,3/2,2,5/2,3];
```

$$R := \left[\frac{1}{2}, 1, \frac{3}{2}, 2, \frac{5}{2}, 3\right]$$

In the following loop the Maple command nops applied to R returns the number of operands in the list R. Hence, it is the number of elements in the list R.

The colon after the loop-ending **od** is crucial. That colon suppresses the output of the whole loop. If it were not there, all the contents of the plot data-structures being generated would print to the screen.

```
• for k from 1 to nops(R) do
    f.k := implicitplot(f = R[k]^2,
    x =-4..4, y =-4..4, scaling=constrained):
  od:
```

Now paste all these individual contours into one plot. Note that again we assign the plot data-structure a name and therefore end the command with a colon to suppress the output of the plot data-structure itself.

```
• p1 := display([f.(1..nops(R))]):
  p1;
```

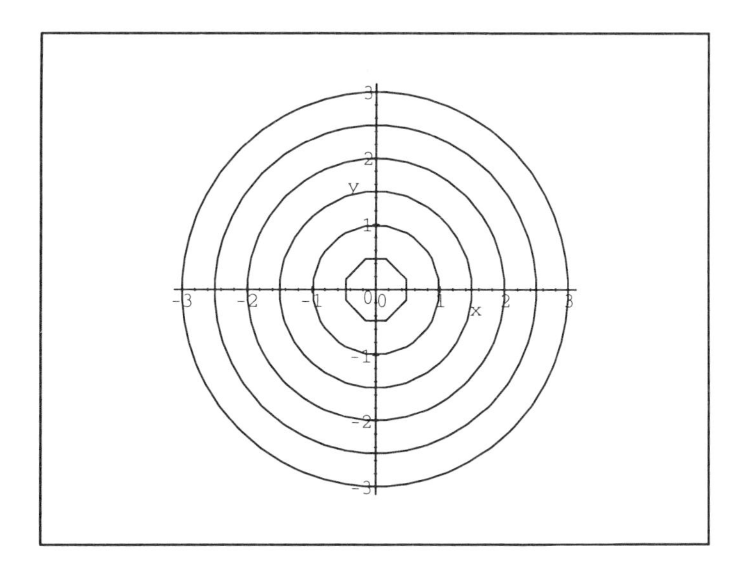

The axes can be "turned off" via the interactive options. In either event, it is against this contour map that we will display the implicit plot of the constraint curve defined by $g(x, y) = 0$.

```
• p2 := implicitplot(g = 0, x = -3..3, y = -3..3):
```

```
display([p1, p2]);
```

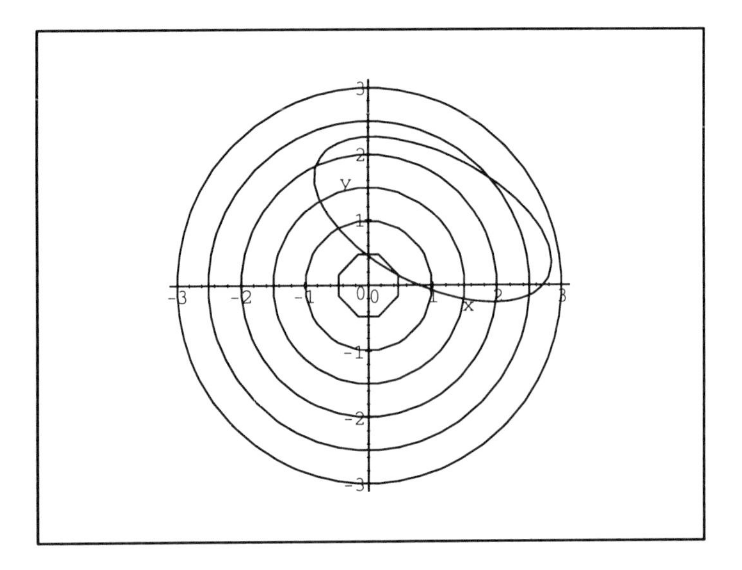

The Lagrange Multiplier theory indicates that the extreme values of $f(x, y)$ on the constraint $g(x, y) = 0$ occur where $g = 0$ is tangent to a level curve of $f(x, y)$.

To find the extreme values of $f(x, y)$ on the constraint $g(x, y) = 0$ form the new objective function

- `F := f - m*g;`

$$F := x^2 + y^2 - m\,(\,17\,y^2 + 12\,x\,y + 8\,x^2 - 46\,y - 28\,x + 17\,)$$

We have used m for the Lagrange Multiplier. The necessary conditions for the extreme point are

- `q1 := diff(F, x);`

$$q1 := 2\,x - m\,(\,12\,y + 16\,x - 28\,)$$

- `q2 := diff(F, y);`

$$q2 := 2\,y - m\,(\,34\,y + 12\,x - 46\,)$$

- `q3 := diff(F, m);`

$$q3 := -17\,y^2 - 12\,x\,y - 8\,x^2 + 46\,y + 28\,x - 17$$

The nonlinear equations that arise in the Lagrange Multiplier method present unique algebraic challenges. Let's find out what Maple can do with these equations.

- `q4 := solve({q1, q2, q3}, {x, y, m});`

$$q4 := \Big\{ m = \%1, x = \frac{56000}{27}\,\%1^3 - \frac{2000}{3}\,\%1^2 + \frac{1196}{27}\,\%1 + \frac{34}{27},$$

$$y = -\frac{8000}{27}\%1^3 + \frac{472}{27}\%1 + \frac{17}{27}\Big\}$$

$$\%1 := \mathrm{RootOf}(\,15200\_Z^2 - 600\_Z - 17 - 100000\_Z^3 + 200000\_Z^4\,)$$

Maple purports to have delivered a solution. The Lagrange Multiplier m is a solution of the fourth-degree polynomial equation contained in the **RootOf** expression. Then, x and y are given in terms of m.

To get Maple to produce the four extreme points, use the **allvalues** command, which acts on the **RootOf** expression. Notice the second argument to **allvalues** . It is a flag that tells Maple to deliver, for each value of m , only the distinct solutions for x and y.

- ```
q5 := allvalues(q4, d);
```

$$q5 := \Big\{m = \%8,$$

$$x = \frac{56000}{27}\%8^3 - \frac{2000}{3}\%8^2 + \frac{367}{54} + \frac{299}{20250}\sqrt{\%2} + \frac{299}{20250}\%6,$$

$$y = -\frac{8000}{27}\%8^3 + \frac{76}{27} + \frac{59}{10125}\sqrt{\%2} + \frac{59}{10125}\%6\Big\}, \Big\{m = \%7,$$

$$x = \frac{56000}{27}\%7^3 - \frac{2000}{3}\%7^2 + \frac{367}{54} + \frac{299}{20250}\sqrt{\%2} - \frac{299}{20250}\%6,$$

$$y = -\frac{8000}{27}\%7^3 + \frac{76}{27} + \frac{59}{10125}\sqrt{\%2} - \frac{59}{10125}\%6\Big\}, \Big\{m = \%5,$$

$$x = \frac{56000}{27}\%5^3 - \frac{2000}{3}\%5^2 + \frac{367}{54} - \frac{299}{20250}\sqrt{\%2} + \frac{299}{20250}\%3,$$

$$y = -\frac{8000}{27}\%5^3 + \frac{76}{27} - \frac{59}{10125}\sqrt{\%2} + \frac{59}{10125}\%3\Big\}, \Big\{m = \%4,$$

$$x = \frac{56000}{27}\%4^3 - \frac{2000}{3}\%4^2 + \frac{367}{54} - \frac{299}{20250}\sqrt{\%2} - \frac{299}{20250}\%3,$$

$$y = -\frac{8000}{27}\%4^3 + \frac{76}{27} - \frac{59}{10125}\sqrt{\%2} - \frac{59}{10125}\%3\Big\}$$

$$\%1 := \frac{6839}{27000000000} + \frac{1}{1000000000}\sqrt{41145}$$

$$\%2 := \frac{26625\,\%1^{1/3} - 2250000\,\%1^{2/3} - 64}{\%1^{1/3}}$$

$$\%3 := \sqrt{2}\Big(\Big(26625\,\%1^{1/3}\sqrt{\%2} + 1125000\sqrt{\%2}\,\%1^{2/3} + 32\sqrt{\%2}$$
$$+ 1265625\,\%1^{1/3}\Big)\Big/\Big(\%1^{1/3}\sqrt{\%2}\Big)\Big)^{1/2}$$

$$\%4 := \frac{1}{8} - \frac{1}{3000}\sqrt{\%2} - \frac{1}{3000}\%3$$

$$\%5 := \frac{1}{8} - \frac{1}{3000}\sqrt{\%2} + \frac{1}{3000}\%3$$

$$\%6 := I\sqrt{2}\Big(\Big(-26625\,\%1^{1/3}\sqrt{\%2} - 1125000\sqrt{\%2}\,\%1^{2/3} - 32\sqrt{\%2}$$
$$+ 1265625\,\%1^{1/3}\Big)\Big/\Big(\%1^{1/3}\sqrt{\%2}\Big)\Big)^{1/2}$$
```

$$\%7 := \frac{1}{8} + \frac{1}{3000}\sqrt{\%2} - \frac{1}{3000}\%6$$
$$\%8 := \frac{1}{8} + \frac{1}{3000}\sqrt{\%2} + \frac{1}{3000}\%6$$

The complexity of these solutions is a warning that even simple problems can embroil us in exceedingly messy computations.

The quickest way to interpret what we were given is to ask for a floating-point version of these solutions. The Maple response $q5$ is an expression sequence of four sets, each set containing three equations defining values for x , y, and m.

```
• q6 := evalf(q5);
```

$$\begin{aligned} q6 := \{&y = 2.281672767 + .0158058952\,I,\\ &x = .1513697247 - .292904803\,I,\\ &m = .1354928933 + .01302764151\,I\}, \{\\ &m = .1354928933 - .01302764151\,I,\\ &x = .1513697247 + .292904803\,I,\\ &y = 2.281672767 - .0158058952\,I\},\\ &\{m = .2475466415, y = .462446550, x = 2.834421247\},\\ &\{y = .307541252, m = -.0185324281, x = .196172634\} \end{aligned}$$

Two solutions are real and two are complex. Let's evaluate the objective function $f(x, y)$ at each of the real solutions.

```
• subs(q6[3], f);
```

$$8.247800617$$

```
• subs(q6[4], f);
```

$$.1330653240$$

These values are consistent with the plot generated above. There are two circles tangent to the ellipse represented by $g(x, y) = 0$. The point of tangency with the smaller circle is the minimum, and the point of tangency with the larger circle is the maximum.

How might we have obtained the solution of the governing equations ourselves? Perhaps solving the first two equations for $x = x(m)$ and $y = y(m)$ might be a starting place. Then, substitution of $x = x(m)$ and $y = y(m)$ into the constraint equation would give a single equation for m, namely $g(x(m), y(m)) = 0$, from which the extreme values would follow.

Let's try that in Maple.

```
• q7 := solve({q1, q2}, {x, y});
```

$$q7 := \left\{y = \frac{m\,(-23 + 100\,m)}{100\,m^2 + 1 - 25\,m}, x = 2\,\frac{m\,(-7 + 50\,m)}{100\,m^2 + 1 - 25\,m}\right\}$$

- `q8 := subs(q7, q3);`

$$q8 := -17\,\frac{m^2\,(\,-23+100\,m\,)^2}{(\,100\,m^2+1-25\,m\,)^2} - 24\,\frac{m^2\,(\,-7+50\,m\,)\,(\,-23+100\,m\,)}{(\,100\,m^2+1-25\,m\,)^2} - 32\,\frac{m^2\,(\,-7+50\,m\,)^2}{(\,100\,m^2+1-25\,m\,)^2} + 46\,\frac{m\,(\,-23+100\,m\,)}{100\,m^2+1-25\,m} + 56\,\frac{m\,(\,-7+50\,m\,)}{100\,m^2+1-25\,m} - 17$$

Equation $q8$ certainly needs simplification. Since there are common denominators to be manipulated, use the Maple command **normal** .

- `q9 := normal(q8);`

$$q9 := \frac{15200\,m^2 - 100000\,m^3 + 200000\,m^4 - 600\,m - 17}{(\,100\,m^2+1-25\,m\,)^2}$$

The numerator of $q9$ is the polynomial contained in Maple's **RootOf** expression above.

The real zeros of $q9$ generate the coordinates of the extreme points (x, y).

- `plot(numer(q9), m = -.02 .. .25);`

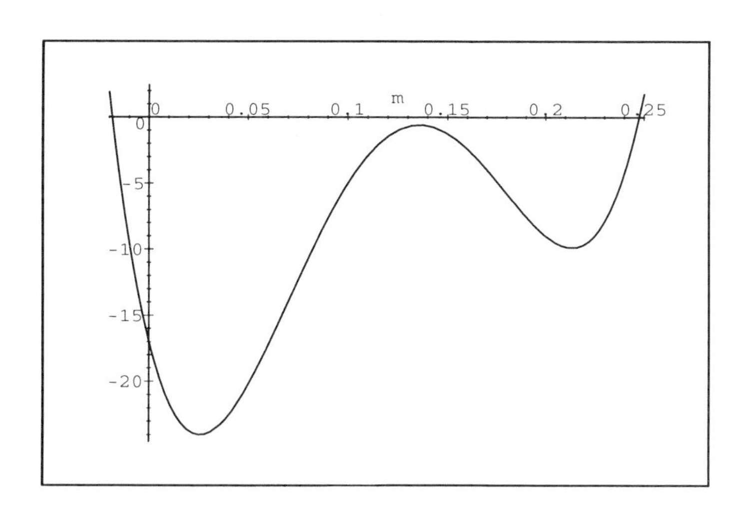

A direct numeric solution for the real values of m:

- `m1 := fsolve(numer(q9), m, m = -1..0);`

$$m1 := -.01853242810$$

- `m2 := fsolve(numer(q9), m, m = 0..1);`

$$m2 := .2475466415$$

From the values of m, we can recover the coordinates of the extreme points.

- `e1 := subs(m = m1, q7);`

$$e1 := \{ y = .3075412507, x = .1961726347 \}$$

- `e2 := subs(m = m2, q7);`

$$e2 := \{ y = .4624465518, x = 2.834421252 \}$$

The labels $e1$ and $e2$ on the extreme points allow us to compute the values of $f(x, y)$ at each.

- `subs(e1, f);`

$$.1330653235$$

- `subs(e2, f);`

$$8.247800647$$

Because the constraint curve defined by $g(x, y) = 0$ is a closed curve, it is interesting to examine how various quantities change as the constraint curve is traversed. This is best done through a continuous parametrization of the curve.

An obvious parametrization of the constraint curve can be obtained by merely solving $g(x, y) = 0$ for $y = y(x)$. However, this results in two branches to the parametrization, making a study of variations on the contour more complicated.

We declare a continuous parametrization and then describe how it was found.

$$\begin{aligned} x &= 1 + \frac{4cos(t) + sin(t)}{\sqrt{5}} \\ y &= 1 - 2\frac{cos(t) - sin(t)}{\sqrt{5}} \end{aligned}$$

Since an ellipse in the standard form

$$(x/a)^2 + (y/b)^2 = 1$$

can be parametrized via $x = a\cos(t), y = b\sin(t)$, we try to put $g(x, y) = 0$ into this standard form. The process starts with a rotation to remove the xy term in $g(x, y)$. The set Q below will contain the transformation equations for a rotation of coordinates about the origin in the xy-plane.

- `Q := { x = cos(s)*X - sin(s)*Y, y = sin(s)*X + cos(s)*Y};`

$$Q := \{ x = \cos(s)X - \sin(s)Y, y = \sin(s)X + \cos(s)Y \}$$

Next, apply this rotation to the ellipse represented by $g(x, y) = 0$.

- ```
  Q1 := subs(Q, g);
  ```

$$Q1 := 17(\sin(s)X + \cos(s)Y)^2 + 12(\cos(s)X - \sin(s)Y)(\sin(s)X + \cos(s)Y) + 8(\cos(s)X - \sin(s)Y)^2 - 46\sin(s)X - 46\cos(s)Y - 28\cos(s)X + 28\sin(s)Y + 17$$

Clear parentheses by

- ```
  Q2 := expand(Q1);
  ```

$$Q2 := 17\sin(s)^2X^2 + 18\sin(s)X\cos(s)Y + 17\cos(s)^2Y^2 + 12\cos(s)X^2\sin(s) + 12\cos(s)^2XY - 12\sin(s)^2YX - 12\sin(s)Y^2\cos(s) + 8\cos(s)^2X^2 + 8\sin(s)^2Y^2 - 46\sin(s)X - 46\cos(s)Y - 28\cos(s)X + 28\sin(s)Y + 17$$

The value of s that eliminates all XY terms is our immediate goal. This requires that we collect all such terms.

First, replace the product $X*Y$ with a single new name. Maple will not do this with its **subs** command, so we use the **powsubs** command, designed for addressing factors as subexpressions. The **powsubs** command is found in the student package.

- ```
  with(student):
  ```
- ```
  Q3 := powsubs(X*Y = r, Q2);
  ```

$$Q3 := 17\sin(s)^2X^2 + 18\sin(s)\cos(s)r + 17\cos(s)^2Y^2 + 12\cos(s)X^2\sin(s) + 12\cos(s)^2r - 12\sin(s)^2r - 12\sin(s)Y^2\cos(s) + 8\cos(s)^2X^2 + 8\sin(s)^2Y^2 - 46\sin(s)X - 46\cos(s)Y - 28\cos(s)X + 28\sin(s)Y + 17$$

The coefficient of r is the term multiplying $X\,Y$. If this term can be extracted and made to be identically zero, we would have removed the evidence, and hence the reality, of rotation.

- ```
  Q4 := coeff(Q3, r);
  ```

$$Q4 := 18\sin(s)\cos(s) + 12\cos(s)^2 - 12\sin(s)^2$$

The angle s through which we must rotate the ellipse to align its axes with the coordinate axes in the XY-plane is found by

- ```
  Q5 := solve(Q4 = 0, s);
  ```

$$Q5 := -\arctan\left(\frac{1}{2}\right), \arctan(2)$$

Thus, there are two angles of rotation that will eliminate the crossterms from g. These differ only in the direction through which the rotation is made. We'll select the first angle and substitute it into the equations defining the rotated coordinates. This will give us the specific realization of the rotated coordinate system in which the ellipse defined by $g(x, y) = 0$ has its axes parallel to the coordinate axes.

- `Q6 := subs(s=Q5[1], Q);`

$$Q6 := \left\{x = \cos\left(-\arctan\left(\frac{1}{2}\right)\right) X - \sin\left(-\arctan\left(\frac{1}{2}\right)\right) Y,\right.$$
$$\left. y = \sin\left(-\arctan\left(\frac{1}{2}\right)\right) X + \cos\left(-\arctan\left(\frac{1}{2}\right)\right) Y\right\}$$

We get Maple to do the required trigonometric simplifications in $Q6$ via

- `Q7 := expand(Q6);`

$$Q7 := \left\{y = -\frac{1}{5}\sqrt{5}\,X + \frac{2}{5}\sqrt{5}\,Y, x = \frac{2}{5}\sqrt{5}\,X + \frac{1}{5}\sqrt{5}\,Y\right\}$$

The set $Q7$ contains the transformation equations relating the original xy-coordinate system to the (rotated) XY-system in which the ellipse has its axes parallel to those in the XY-plane. The next step is to make this coordinate change in $g(x, y)$ to obtain

- `Q8 := subs(Q7, g);`

$$Q8 := 17\left(-\frac{1}{5}\sqrt{5}\,X + \frac{2}{5}\sqrt{5}\,Y\right)^2$$
$$+ 12\left(\frac{2}{5}\sqrt{5}\,X + \frac{1}{5}\sqrt{5}\,Y\right)\left(-\frac{1}{5}\sqrt{5}\,X + \frac{2}{5}\sqrt{5}\,Y\right)$$
$$+ 8\left(\frac{2}{5}\sqrt{5}\,X + \frac{1}{5}\sqrt{5}\,Y\right)^2 - 2\sqrt{5}\,X - 24\sqrt{5}\,Y + 17$$

There are still parentheses to be cleared, and this can be done via the **expand** command.

- `Q9 := expand(Q8);`

$$Q9 := 5\,X^2 + 20\,Y^2 - 2\sqrt{5}\,X - 24\sqrt{5}\,Y + 17$$

The equation represented by setting $Q9 = 0$ is the equation of the ellipse $g(x, y) = 0$ rotated so that its semi-major and semi-minor axes are parallel to the coordinate axes in the XY-frame.

The rotation initially took place about the origin of the xy-plane and the center of the ellipse in the XY-system is not at the origin of the XY-plane. To find the center of the ellipse in the XY-system, we will have to complete the square in both X and Y.

- `Q10 := completesquare(Q9, [X, Y]);`

$$Q10 := 20\left(Y - \frac{3}{5}\sqrt{5}\right)^2 - 20 + 5\left(X - \frac{1}{5}\sqrt{5}\right)^2$$

Thus, $Q10 = 0$ almost gives us the constraint ellipse in the standard form

$$(1/4)(X-h)^2 + (Y-k)^2 = 1$$

where $(h,k) = (1/\sqrt{5}, 3/\sqrt{5})$. But we can see that the parametrization we have been seeking is given by

$$\begin{aligned} X &= h + 2\cos(t) \\ Y &= k + \sin(t) \end{aligned}$$

But this parametrization needs to be expressed in terms of x and y. The equations in $Q7$ define the transformation $x = x(X,Y), y = y(X,Y)$. We need to invert the tranformation so it has the form $X = X(x,y), Y = Y(x,y)$. Thus,

- `Q11 := solve(Q7, {X, Y});`

$$Q11 := \left\{X = \frac{1}{5}\sqrt{5}(-y+2x), Y = \frac{1}{5}(x+2y)\sqrt{5}\right\}$$

If we equate $X = h + 2\cos(t)$ with $X = X(x,y)$ from $Q11$, and similarly if we equate $Y = k + sin(t)$ with $Y = Y(x,y)$ from $Q11$, we have equations of the form

$$\begin{aligned} X &= h + 2\cos(t) &&= X(x,y) \\ Y &= k + \sin(t) &&= Y(x,y) \end{aligned}$$

from which we can obtain x and y in terms of just $\cos(t)$ and $\sin(t)$. That will be the desired parametrization. Thus,

- `Q12 := subs(X = 1/sqrt(5) + 2*cos(t), Y = 3/sqrt(5) + sin(t), Q11);`

$$Q12 := \left\{ \frac{1}{5}\sqrt{5} + 2\cos(t) = \frac{1}{5}\sqrt{5}(-y+2x), \frac{3}{5}\sqrt{5} + \sin(t) = \frac{1}{5}(x+2y)\sqrt{5} \right\}$$

Solving the equations in $Q12$ for $x = x(t)$ and $y = y(t)$ produces the promised parametrization of the original ellipse.

- `Q13 := solve(Q12, {x, y});`

$$Q13 := \left\{x = 1 + \frac{4}{5}\sqrt{5}\cos(t) + \frac{1}{5}\sqrt{5}\sin(t), \; y = 1 + \frac{2}{5}\sqrt{5}\sin(t) - \frac{2}{5}\sqrt{5}\cos(t)\right\}$$

The first use of this parametrization is a study of the variation of the values of $f(x, y)$ as the ellipse is traversed.

- `Q14 := subs(Q13, f);`

$$Q14 := \left(1+\frac{4}{5}\sqrt{5}\cos(t)+\frac{1}{5}\sqrt{5}\sin(t)\right)^2 + \left(1+\frac{2}{5}\sqrt{5}\sin(t)-\frac{2}{5}\sqrt{5}\cos(t)\right)^2$$

- `plot(Q14, t = 0..2*Pi);`

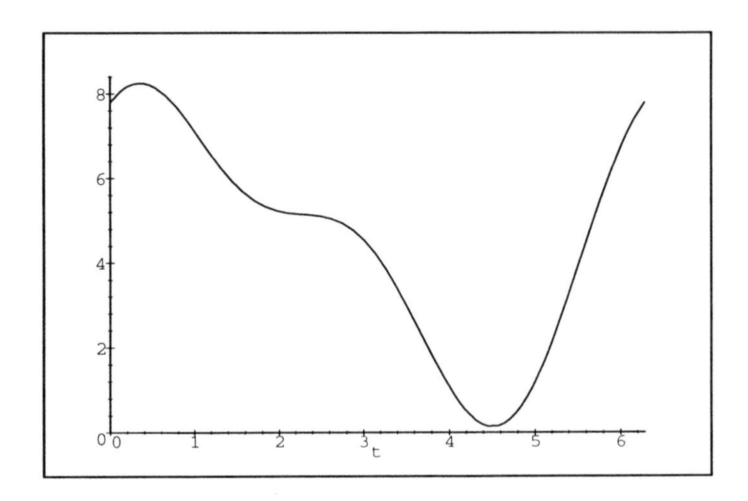

The real maximum appears to occur in the subinterval $(0, 1)$ and the real minimum, in $(4, 5)$.

A computation verifies this.

- `Q15 := diff(Q14, t);`

$$Q15 := 2\left(1+\frac{4}{5}\sqrt{5}\cos(t)+\frac{1}{5}\sqrt{5}\sin(t)\right)\left(-\frac{4}{5}\sqrt{5}\sin(t)+\frac{1}{5}\sqrt{5}\cos(t)\right) + 2\left(1+\frac{2}{5}\sqrt{5}\sin(t)-\frac{2}{5}\sqrt{5}\cos(t)\right)\left(\frac{2}{5}\sqrt{5}\cos(t)+\frac{2}{5}\sqrt{5}\sin(t)\right)$$

- `t1 := fsolve(Q15, t, t = 0..1);`

$$t1 := .3464657569$$

- `t2 := fsolve(Q15, t, t = 4..5);`

$$t2 := 4.506288940$$

Substituting these angles into the equations contained in the set $Q13$ will give us the x- and y-coordinates corresponding to the extreme values of $f(x, y)$.

- `evalf(subs(t = t1, Q13));`

$$\{y = .4624465514, x = 2.834421249\}$$

- `evalf(subs(t = t2, Q13));`

$$\{y = .3075412511, x = .1961726338\}$$

We have reproduced the two extreme points found earlier.

Next, let's examine how the slopes along the constraint ellipse vary. Remember: at the extreme points the slopes on the ellipse and on the level curve of $f(x, y)$ are equal.

The easiest way to get the slopes for both the ellipse and the contours of $f(x, y)$ will be through implicit differentiation.

- `w1 := subs(y = y(x), f);`

$$w1 := x^2 + \mathrm{y}(x)^2$$

- `w2 := diff(w1, x);`

$$w2 := 2\,x + 2\,\mathrm{y}(x)\left(\frac{\partial}{\partial x}\mathrm{y}(x)\right)$$

- `w3 := solve(w2, diff(y(x), x));`

$$w3 := -\frac{x}{\mathrm{y}(x)}$$

- `w4 := subs(y(x) = y, w3);`

$$w4 := -\frac{x}{y}$$

We repeat this calculation for $g(x, y) = 0$.

- `w5 := subs(y = y(x), g);`

$$w5 := 17\,\mathrm{y}(x)^2 + 12\,x\,\mathrm{y}(x) + 8\,x^2 - 46\,\mathrm{y}(x) - 28\,x + 17$$

- `w6 := diff(w5, x);`

$$w6 := 34\,\mathrm{y}(x)\left(\frac{\partial}{\partial x}\mathrm{y}(x)\right) + 12\,\mathrm{y}(x) + 12\,x\left(\frac{\partial}{\partial x}\mathrm{y}(x)\right) + 16\,x - 46\left(\frac{\partial}{\partial x}\mathrm{y}(x)\right) - 28$$

- `w7 := solve(w6, diff(y(x), x));`

$$w7 := -\frac{12\,\mathrm{y}(x) + 16\,x - 28}{34\,\mathrm{y}(x) + 12\,x - 46}$$

- `w8 := subs(y(x) = y, w7);`

$$w8 := -\frac{12\,y + 16\,x - 28}{34\,y + 12\,x - 46}$$

Next, we parametrize, along the ellipse, each of these expressions for slopes.

- `w9 := subs(Q13, w4);`

$$w9 := -\frac{1+\frac{4}{5}\sqrt{5}\cos(t)+\frac{1}{5}\sqrt{5}\sin(t)}{1+\frac{2}{5}\sqrt{5}\sin(t)-\frac{2}{5}\sqrt{5}\cos(t)}$$

- `w10 := subs(Q13, w8);`

$$w10 := -\frac{8\sqrt{5}\sin(t)+8\sqrt{5}\cos(t)}{16\sqrt{5}\sin(t)-4\sqrt{5}\cos(t)}$$

Clearly, we want to plot these formulas.

- `plot(w9, t = 0..2*Pi, 'y'' = -10..10);`

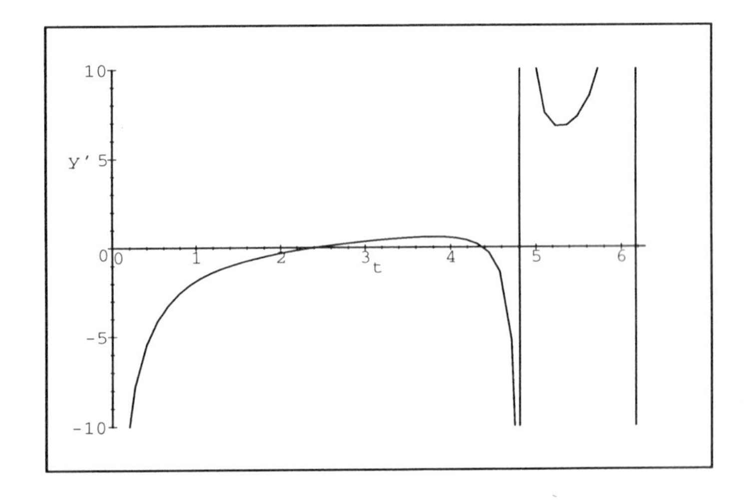

- `plot(w10, t = 0..2*Pi, 'y'' = -10..10);`

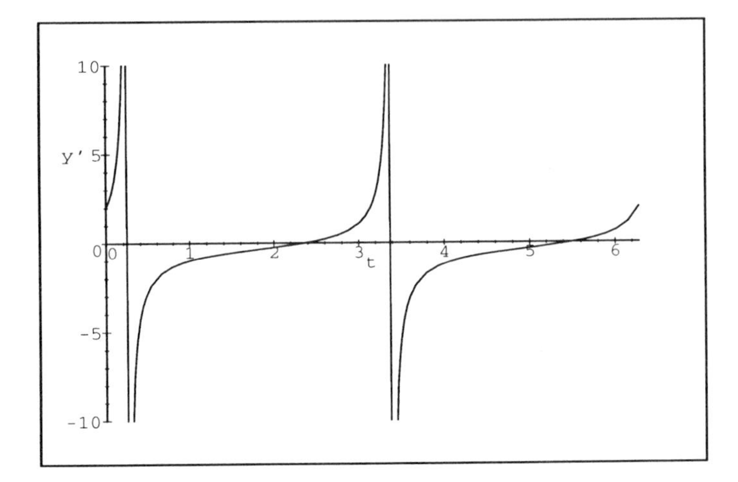

Can we determine where these slopes are equal?

First, try plotting both curves together.

- `plot({w9, w10}, t = 0..2*Pi, 'y'' = -10..10);`

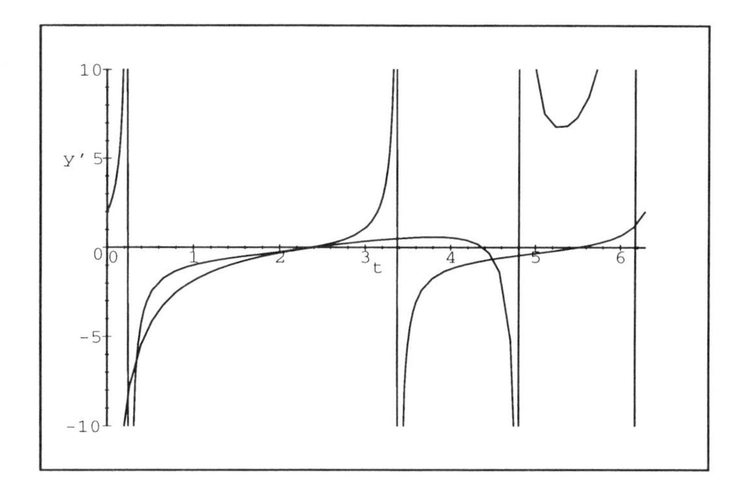

The figure seems to indicate three places where the slopes are equal. However, a closer look at the subinterval (2,3) is revealing.

- `plot({w9,w10}, t = 2..3);`

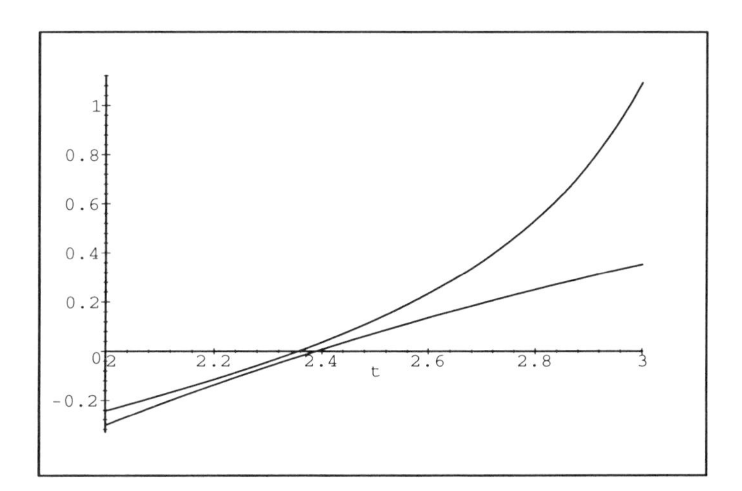

The curves do not intersect in this subinterval. Hence, there are only two values of t for which the slopes are equal. We compute these values by subtracting the two expressions for the derivatives and simplifying to a single fraction. The numerator of this fraction, when set equal to zero, gives the equation to be solved for points where slopes are equal.

- `w11 := normal(w9 - w10);`

$$w11 := -5\,\frac{2\sin(t) - 3\cos(t) + 3\sqrt{5}\cos(t)\sin(t)}{\left(5 + 2\sqrt{5}\sin(t) - 2\sqrt{5}\cos(t)\right)\,(4\sin(t) - \cos(t))}$$

- `w12 := numer(w11);`

$$w12 := -10\sin(t) + 15\cos(t) - 15\sqrt{5}\cos(t)\sin(t)$$

This equation ($w12 = 0$) tells us where the slope on the constraint ellipse equals the slope on the level curve of $f(x, y)$. A picture is worth a thousand words, so why not look at the graph of the function in $w12$.

- `plot(w12, t = 0..2*Pi);`

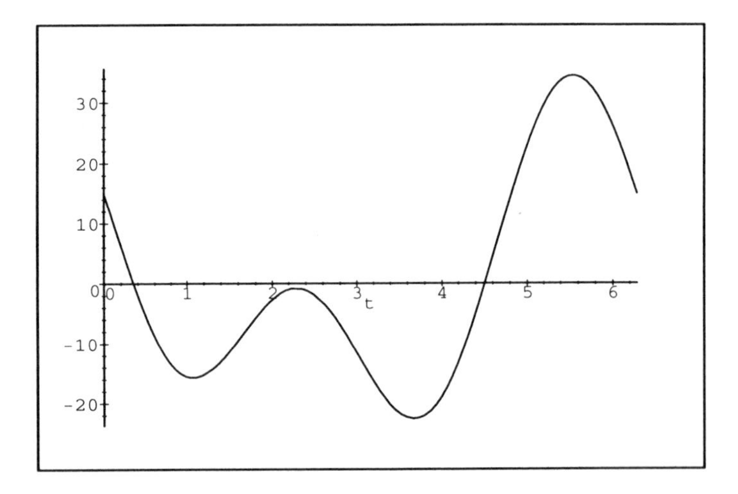

There appear to be two roots, one in the interval $[0, 1]$ and one in the interval $[4, 5]$. Perhaps the simplest way to determine what these roots are is via

- `w13 := fsolve(w12, t, t = 0..1);`

$$w13 := .3464657569$$

- `w14 := fsolve(w12, t, t = 4..5);`

$$w14 := 4.506288940$$

These two roots agree with the values of t computed earlier as $t1$ and $t2$. We have again determined the same two points.

Our next study will be an examination of how gradients vary as we march around the ellipse. Some commands in the linear algebra package will be useful.

- `with(linalg):`

```
Warning: new definition for   norm
Warning: new definition for   trace
```

- `gradf := grad(f, [x, y]);`

$$gradf := [2x \;\; 2y]$$

```
• gradg := grad(g, [x, y]);
```

$$gradg := [\,12\,y + 16\,x - 28\;\; 34\,y + 12\,x - 46\,]$$

Once again we caution that these gradient vectors are *column* vectors, even though Maple insists on printing them horizontally. Successful use of the linear algebra package demands that the user become comfortable with this quirk.

We will explore these gradients along the constraint ellipse. Hence, we want to parametrize them along this ellipse. Note how the appropriate substitutions are made into the operands of the gradient vectors.

```
• gradft := subs(Q13, op(gradf));
```

$$gradft := \left[2 + \frac{8}{5}\sqrt{5}\cos(t) + \frac{2}{5}\sqrt{5}\sin(t)\;\; 2 + \frac{4}{5}\sqrt{5}\sin(t) - \frac{4}{5}\sqrt{5}\cos(t)\right]$$

```
• gradgt := subs(Q13, op(gradg));
```

$$gradgt := \left[8\sqrt{5}\sin(t) + 8\sqrt{5}\cos(t)\;\; 16\sqrt{5}\sin(t) - 4\sqrt{5}\cos(t)\right]$$

Let's examine how the angle between these gradients varies along the constraint ellipse.

```
• plot(angle(gradft, gradgt), t = 0..2*Pi);
```

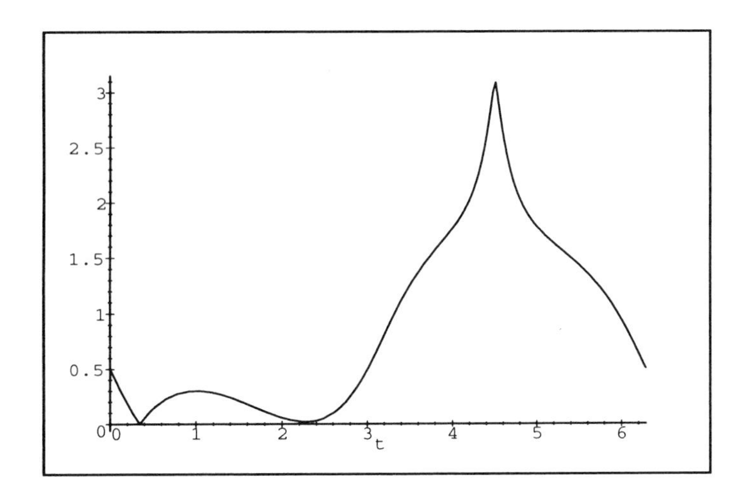

The angle between the gradients seems to be zero in the intervals $(0, 1)$ and $(2, 3)$. However, a closer look in the interval $(2, 3)$ shows otherwise.

```
• plot(angle(gradft, gradgt), t = 2..3);
```

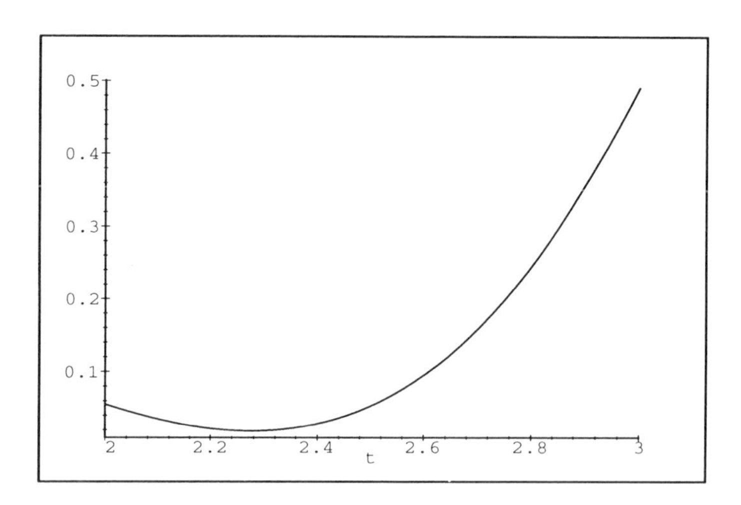

That the angle between the gradients is zero in $(0, 1)$ is consistent with our previously calculated value $t1 = .3464657572$. At first glance it is strange that there is no other place where this angle is again zero.

However, preconceptions are always risky, and here the preconception that the angle between the gradients is zero twice is wrong. The angle is once zero and once π, indicating that at the maximum the gradients point in the same direction, but at the minimum the gradients are collinear but point in opposite directions.

We complete our investigations by drawing a figure that displays scaled gradient vectors at points along the constraint ellipse. We will normalize gradients of g to have length one, and we will scale gradients of f to have length $1/2$.

The strategy will be to marshal appropriate analytic formulas and then evaluate these formulas at specific points on the ellipse. The formulas are best understood via a diagram. Execution of the following Maple code will generate the Auxilliary Figure (below), on which we base an explanation of the remaining computations.

- ```
  pp1:=plot([[0,0],[2.3,1.3]]):
  pp2:=plot([[2.3,1.3],[2.9,2]]):
  pp3:=plot([[0,0],[2.9,2]]):
  pp4:=textplot({[2.5,1.3,'a'],[3,2,'b']}):
  display([p2,pp1,pp2,pp3,pp4],scaling=constrained);
  ```
  ```

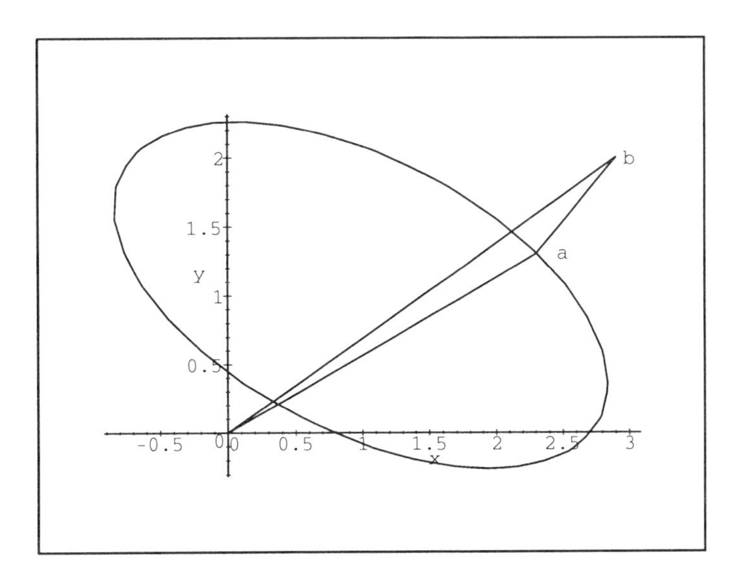

Auxilliary Figure

Point a is on the ellipse and segment ab represents a gradient vector with tail at point a. Computing the components of this gradient vector does not immediately give us the coordinates of point b at the tip of the gradient. We must first obtain the components of the vector from the origin, O, to point b, since the components of this vector are the coordinates of point b. The vector Ob is found by the vector sum of the vectors Oa and ab.

We begin the task of obtaining the appropriate formulas by first extracting expressions for $x(t)$ and $y(t)$, coordinates of points on the ellipse. Remember, these expressions are stored in the set $Q13$ and expressions for the gradients have been computed above and stored in the variables $gradft$ and $gradgt$.

- `u := subs(Q13, x);`

$$u := 1 + \frac{4}{5}\sqrt{5}\cos(t) + \frac{1}{5}\sqrt{5}\sin(t)$$

- `v := subs(Q13, y);`

$$v := 1 + \frac{2}{5}\sqrt{5}\sin(t) - \frac{2}{5}\sqrt{5}\cos(t)$$

To scale the vectors $grad(f)$ and $grad(g)$, we use the **norm** command from the linear algebra package. The simplest way to get Maple to do arithmetic on vectors is to enclose that arithmetic with the **evalm** (i.e., evaluate matrix) command. Thus,

- ```
gradftn := evalm(gradft/norm(gradft,2)/2):
gradgtn := evalm(gradgt/norm(gradgt,2)):
```

Next, we obtain a vector from the origin to the tip of each gradient. This is done by vector addition, as explained earlier. The vector whose components are $u$ and $v$ is actually an arrow from the origin to the point $(u, v)$ on the ellipse. It would be the vector $Oa$ on the
```

Auxilliary Figure. The gradients are the vectors ab in the Auxilliary Figure. Hence, their vector sum is an arrow from the origin to the tip of the gradient. This was called *Ob* in the Auxilliary Figure.

```
• tipf := evalm(vector([u, v]) + gradftn):
  tipg := evalm(vector([u, v]) + gradgtn):
```

At some 30 points around the ellipse, we want the coordinates of the tails of gradient vectors. We also want the coordinates of the corresponding tips of these gradients. Let's start with a list of 30 values of the parameter that generates the ellipse.

```
• T := [seq(evalf(2*Pi*k/30), k = 0..29)]:
```

The corresponding list of points on the ellipse would therefore be

```
• B := [seq(evalf(subs(t = T[k], [u, v])), k = 1..30)]:
```

Next, we get lists of the vectors from the origin to the tips of the scaled gradients at these 30 points on the ellipse. We used vector addition to obtain the components of the vectors from the origin to the tips of the scaled gradients. We will then be able to convert these components into coordinates.

```
• Af := [seq(evalf(subs(t = T[k], op(tipf))), k = 1..30)]:
  Ag := [seq(evalf(subs(t = T[k], op(tipg))), k = 1..30)]:
```

The vectors from the origin to the tips of the scaled gradients have served their purpose. They will now be converted to lists of coordinates of the points at the tips of the scaled gradients.

```
• AAf := map(convert, Af, list):
  AAg := map(convert, Ag, list):
```

Then, we create sequences of coordinate pairs for the tail and tip of each scaled gradient to be drawn.

```
• setf := seq([B[k], AAf[k]], k = 1..30):
  setg := seq([B[k], AAg[k]], k = 1..30):
```

Finally, we plot the line segments connecting the tips and tails of the scaled gradients, along with a parametric plot of the ellipse.

```
• plot({setf, setg, [u, v, t = 0..2*Pi]});
```

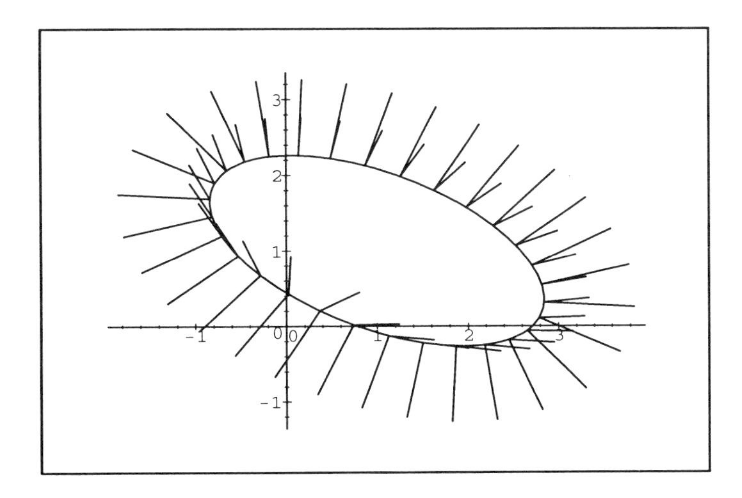

And a view with the axes removed.

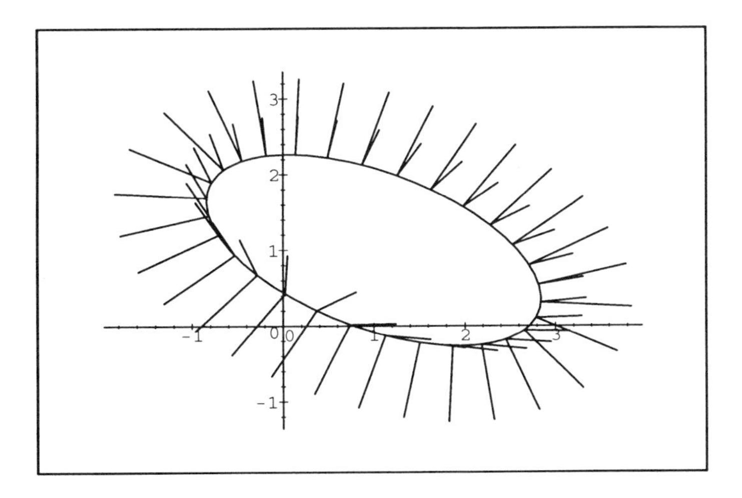

From either figure we can see how the gradients are related as the constraining ellipse is traversed. Remember, the longer line segments represent the gradients of g, and the shorter ones, f. If the lower-right vertex of the ellipse is the starting point, then motion counterclockwise on the ellipse sees the gradients becoming collinear. Along most of the top portion of the ellipse, $grad(f)$ then lies to the right of $grad(g)$. It appears that the gradients again become collinear just before the left vertex, but this appearance is a function of the graphics only. The gradients approach very closely but do not coincide. In fact, further motion counterclockwise has the gradients again oriented the same way, with $grad(f)$ to the right of $grad(g)$.

It is on the lower portion of the ellipse that the gradients become collinear, but this time they are pointing in opposite directions.

Unit 28: The Lagrange Multiplier, Part Three

This is an example of constrained optimization taken from an economics course.

Find the extreme values of the linear function $f(x1, x2) = w1\ x1 + w2\ x2$ subject to the constraint relationship $y = x1^{a1}\ x2^{a2}$.

It is useful to point out that the variables are $x1$ and $x2$. The quantities y , $w1$, $w2$, $a1$, and $a2$ are all parameters. In particular, the constraint equation intends y to be "fixed" so that the constraint implicitly defines either $x1 = x1(x2)$ or $x2 = x2(x1)$.

Here is a Maple solution for this problem. First, enter the objective function, giving it the name F. Use L as the Lagrange Multiplier.

```
• F := w1*x1 + w2*x2 + L*(y - x1^a1 * x2^a2);
```

$$F := w1\ x1 + w2\ x2 + L\,(y - x1^{a1}\ x2^{a2})$$

Calculate and set equal to zero the partial derivatives with respect to $x1$, $x2$, and L. This third equation is just the constraint itself. The names $e1$, $e2$, and $e3$ are attached to these three resulting equations.

```
• e1 := diff(F, x1) = 0;
```

$$e1 := w1 - \frac{L\ x1^{a1}\ a1\ x2^{a2}}{x1} = 0$$

```
• e2 := diff(F, x2) = 0;
```

$$e2 := w2 - \frac{L\ x1^{a1}\ x2^{a2}\ a2}{x2} = 0$$

```
• e3 := diff(F, L ) = 0;
```

$$e3 := y - x1^{a1}\ x2^{a2} = 0$$

First, let's see if Maple will solve these three nonlinear equations directly.

```
• solve({e1, e2, e3}, {x1, x2, L});
```

Maple does not solve this system. Nonlinear equations are generally difficult, especially when there are only symbols and no numbers. What complicates these equations for Maple is the prevalence of exponents. Were we solving this system "by-hand," we'd probably take logarithms to unravel the exponents. To try that approach in Maple, just as in a "by-hand" calculation, we'll need left- and right-hand sides to which we can apply **ln** .

Load the student package to gain access to the **isolate** command. The difference between **isolate** and **solve** is that **isolate** creates an equation with the "solved for" item on the left.

Solve tends to deliver just the solution value.

```
• with(student):
  q1 := isolate(e1, w1);
```

$$q1 := w1 = \frac{L\ x1^{a1}\ a1\ x2^{a2}}{x1}$$

- `q2 := isolate(e2, w2);`

$$q2 := w2 = \frac{L\, x1^{a1}\, x2^{a2}\, a2}{x2}$$

- `q3 := isolate(e3, y);`

$$q3 := y = x1^{a1}\, x2^{a2}$$

Next, take logarithms of both sides of the equations $q1$, $q2$, and $q3$. Since we will have to simplify after taking logarithms, and since it will take the **map** command to apply the logarithm to both sides of each equation, we compose (using the "at" symbol @) the **simplify** and **ln** functions in the act of applying them to both sides.

- `qq1 := map(simplify@ln, q1);`

$$qq1 := \ln(w1) = \ln(L) + a1\ln(x1) - \ln(x1) + \ln(a1) + a2\ln(x2)$$

- `qq2 := map(simplify@ln, q2);`

$$qq2 := \ln(w2) = \ln(L) + a1\ln(x1) + a2\ln(x2) - \ln(x2) + \ln(a2)$$

- `qq3 := map(simplify@ln, q3);`

$$qq3 := \ln(y) = a1\ln(x1) + a2\ln(x2)$$

Let's see if Maple is ready to solve the equations now.

- `q := solve({qq1, qq2, qq3}, {x1, x2, L});`

$$q := \left\{ x1 = e^{\left(-\frac{\ln(w1)\,a2 - \ln(a1)\,a2 - \ln(w2)\,a2 + \ln(a2)\,a2 - \ln(y)}{a2+a1}\right)},\right.$$
$$x2 = e^{\left(\frac{\ln(w1)\,a1 - \ln(a1)\,a1 - a1\ln(w2) + a1\ln(a2) + \ln(y)}{a2+a1}\right)},$$
$$\left. L = e^{\left(-\frac{-\ln(w2)\,a2 + \ln(a2)\,a2 - \ln(y) - \ln(w1)\,a1 + \ln(a1)\,a1 + \ln(y)\,a2 + \ln(y)\,a1}{a2+a1}\right)} \right\}$$

Since there are terms of the form $\exp(\ln(\bullet))$ a Maple **expand** will probably work to simplify the solutions in set q .

- `Q := expand(q);`

$$Q := \left\{ x1 = \frac{a1^{\left(\frac{a2}{a2+a1}\right)}\, w2^{\left(\frac{a2}{a2+a1}\right)}\, y^{\left(\frac{1}{a2+a1}\right)}}{w1^{\left(\frac{a2}{a2+a1}\right)}\, a2^{\left(\frac{a2}{a2+a1}\right)}},\right.$$
$$x2 = \frac{w1^{\left(\frac{a1}{a2+a1}\right)}\, a2^{\left(\frac{a1}{a2+a1}\right)}\, y^{\left(\frac{1}{a2+a1}\right)}}{a1^{\left(\frac{a1}{a2+a1}\right)}\, w2^{\left(\frac{a1}{a2+a1}\right)}},$$
$$\left. L = \frac{w2^{\left(\frac{a2}{a2+a1}\right)}\, y^{\left(\frac{1}{a2+a1}\right)}\, w1^{\left(\frac{a1}{a2+a1}\right)}}{a2^{\left(\frac{a2}{a2+a1}\right)}\, a1^{\left(\frac{a1}{a2+a1}\right)}\, y^{\left(\frac{a2}{a2+a1}\right)}\, y^{\left(\frac{a1}{a2+a1}\right)}} \right\}$$

As anticipated, the exponentials and logarithms are gone. There is a certain amount of simplification yet to be done, but in essence we have solved this problem.

A careful inspection of the expressions for $x1$ and $x2$ in the set Q shows that the combination

$$R = \frac{w1\ a2}{w2\ a1}$$

appears in both. Perhaps $x1$ and $x2$ should be given in terms of R. First, let's extract just $x1$ and $x2$ from the set Q.

```
• Q1 := x1 = subs(Q, x1), x2 = subs(Q, x2);
```

$$Q1 := x1 = \frac{a1^{\left(\frac{a2}{a2+a1}\right)}\, w2^{\left(\frac{a2}{a2+a1}\right)}\, y^{\left(\frac{1}{a2+a1}\right)}}{w1^{\left(\frac{a2}{a2+a1}\right)}\, a2^{\left(\frac{a2}{a2+a1}\right)}},$$
$$x2 = \frac{w1^{\left(\frac{a1}{a2+a1}\right)}\, a2^{\left(\frac{a1}{a2+a1}\right)}\, y^{\left(\frac{1}{a2+a1}\right)}}{a1^{\left(\frac{a1}{a2+a1}\right)}\, w2^{\left(\frac{a1}{a2+a1}\right)}}$$

Here, we introduce the combination represented by R.

```
• Q2 := subs(w1 = R*a1*w2/a2, [Q1]);
```

$$Q2 := \left[x1 = \frac{a1^{\left(\frac{a2}{a2+a1}\right)}\, w2^{\left(\frac{a2}{a2+a1}\right)}\, y^{\left(\frac{1}{a2+a1}\right)}}{\left(\frac{R\,a1\ w2}{a2}\right)^{\left(\frac{a2}{a2+a1}\right)}\, a2^{\left(\frac{a2}{a2+a1}\right)}},\right.$$
$$\left. x2 = \frac{\left(\frac{R\,a1\ w2}{a2}\right)^{\left(\frac{a1}{a2+a1}\right)}\, a2^{\left(\frac{a1}{a2+a1}\right)}\, y^{\left(\frac{1}{a2+a1}\right)}}{a1^{\left(\frac{a1}{a2+a1}\right)}\, w2^{\left(\frac{a1}{a2+a1}\right)}}\right]$$

Once more, clean up the expressions; this time, **simplify** does the job.

```
• simplify(Q2);
```

$$\left[x1 = \left(\frac{R\,a1\ w2}{a2}\right)^{\left(-\frac{a2}{a2+a1}\right)} a1^{\left(\frac{a2}{a2+a1}\right)}\, w2^{\left(\frac{a2}{a2+a1}\right)}\, a2^{\left(-\frac{a2}{a2+a1}\right)}\, y^{\left(\frac{1}{a2+a1}\right)},\right.$$
$$\left. x2 = \left(\frac{R\,a1\ w2}{a2}\right)^{\left(\frac{a1}{a2+a1}\right)} a1^{\left(-\frac{a1}{a2+a1}\right)}\, w2^{\left(-\frac{a1}{a2+a1}\right)}\, a2^{\left(\frac{a1}{a2+a1}\right)}\, y^{\left(\frac{1}{a2+a1}\right)}\right]$$

We leave the reader the task of manipulating the form of L and expressing $x1$ and $x2$ in terms of $w1$, $w2$, $a1$ and $a2$, instead of in terms of R.

Unit 29: Iterated Integration

The lessons learned about the meaning of definite integration as area under a curve can be repeated as an introduction to multiple integration. However, there is no generalization of the **leftbox** and **rightbox** commands of the student package. Implementating multiple integrals via the definitions requires the more basic approach of increments and sums. However, experience acquired with single integrals does set the pattern for multiple integration, and the directness and clarity achieved in the one-variable case does carry over into higher dimensions.

We illustrate this approach for double integrals, leaving the reader to implement the case of triple integration. Finally, it is clear that we are computing iterated integrals and not double integrals.

As our working example let's take the function

```
• f := x^2 + y^2;
```

$$f := x^2 + y^2$$

and find the volume under it on the square for which both x and y are in $[0, 1]$. We begin by attempting to construct a figure similar to that produced by the **leftbox** or **rightbox** commands. For this, we need items from both the linear algebra and the plots packages.

```
• with(linalg):

Warning: new definition for   norm
Warning: new definition for   trace

• with(plots):
```

A plot showing parallelepipeds that approximate the volume under a surface can be created by the following command.

```
• matrixplot(matrix(10,10,(i,j)->i^2+j^2),
      heights=histogram, axes=boxed);
```

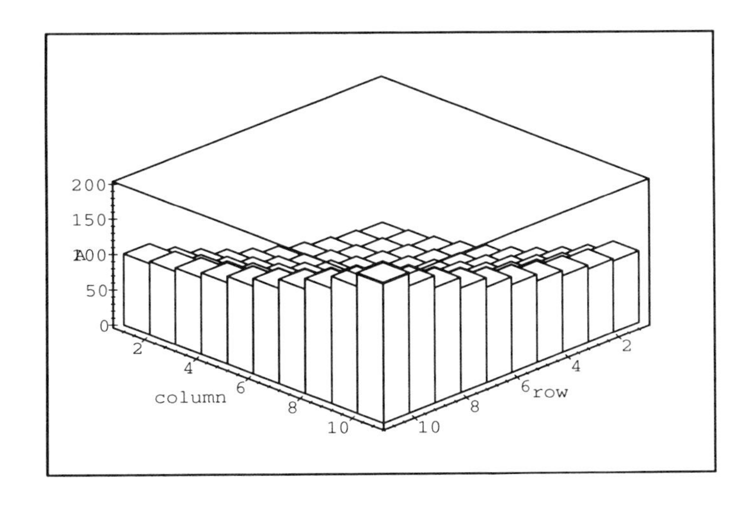

This figure illustrates the basic notions about volume under a surface being approximated by a collection of parallelepipeds (blocks) with rectangular bases and heights determined by the surface. In this configuration the generalization of **leftbox** to "lower-left box" is seen. In each direction the height of the function is determined at the "lower" end of the subinterval for that variable. That puts the evaluation point in the "back left" corner of the rectangular base of each block.

Analytically, the corresponding sums are created by defining increments along each axis and by defining uniform partition points along each axis. The inert sums and limits are the obvious generalizations of the single-variable case.

First, we define increments.

- `hx := 1/n;`

$$hx := \frac{1}{n}$$

- `hy := 1/m;`

$$hy := \frac{1}{m}$$

Then, we define grid points.

- `xj := hx * j;`

$$xj := \frac{j}{n}$$

- `yk := hy * k;`

$$yk := \frac{k}{m}$$

Next, obtain the function values at the grid points.

- `fjk := subs(x = xj, y = yk, f);`

$$fjk := \frac{j^2}{n^2} + \frac{k^2}{m^2}$$

Now, form the approximating double sum using **Sum**, the inert form of the **sum** command.

- `q := Sum(Sum(fjk * hx * hy, j = 0..n), k = 0..m);`

$$q := \sum_{k=0}^{m} \left(\sum_{j=0}^{n} \frac{\frac{j^2}{n^2} + \frac{k^2}{m^2}}{n\,m} \right)$$

Also, form the iterated limit using **Limit**, the inert form of the **limit** command.

```
• q1 := Limit(Limit(q, n = infinity), m = infinity);
```

$$q1 := \lim_{m\to\infty} \lim_{n\to\infty} \sum_{k=0}^{m} \left(\sum_{j=0}^{n} \frac{\frac{j^2}{n^2} + \frac{k^2}{m^2}}{n\,m} \right)$$

The value of this sum is

```
• value(q1);
```

$$\frac{2}{3}$$

If we now introduce the iterated definite integral we can check this result via direct integration.

```
• q2 := Int(Int(f,x = 0..1), y = 0..1);
```

$$q2 := \int_0^1 \int_0^1 x^2 + y^2 \, dx \, dy$$

```
• value(q2);
```

$$\frac{2}{3}$$

As happy as I am with this agreement, I really wish Maple would put parentheses around the integrand.

Interestingly enough, the student package contains a **Doubleint** command that will create an appropriate iterated integral.

```
• with(student):
  q3 := Doubleint(f, x = 0..1, y = 0..1);
```

$$q3 := \int_0^1 \int_0^1 x^2 + y^2 \, dx \, dy$$

Obviously, the **Doubleint** command is equivalent to iterating two inert integration (**Int**) commands. The **Doubleint** command is smart enough to handle an inner integral with non-constant bounds. For example,

```
• q4 := Doubleint(f, x = y^2..y, y = 0..1);
```

$$q4 := \int_0^1 \int_{y^2}^{y} x^2 + y^2 \, dx \, dy$$

```
• value(q4);
```

$$\frac{3}{35}$$

We can compare this "black-box" result with that given by iterating the **int** command.

```
• int(int(f, x = y^2..y), y = 0..1);
```

$$\frac{3}{35}$$

It is unfortunate that the **changevar** command (for changing variables in an integral) cannot complete a multivariate change of variables in a double integral.

```
• changevar({x = r*cos(t), y = r*sin(t)}, q4, [r, t]);

Warning: Computation of new ranges not implemented
```

$$\int\int r^2\ |r|\ dr\ dt$$

Maple can correctly transform the integrand but it cannot analyze the domain over which the integration is to take place. If you think about how we usually do this, you will remember that we use a highly geometric approach. We draw the region of integration in cartesian coordinates and then describe that region in the new (here, polar) coordinates. There is no analytic algorithm for rewriting the domain of integration. Perhaps it is from this difficulty that Maple rightly shrinks.

Since the techniques illustrated here generalize to triple integrals, the reader is left the pleasure of experimenting with that case.

Finally, there is some code in the Maple Share Library (available by ftp access at several computers worldwide, including `ftp.maplesoft.on.ca`) for sketching the curve of intersection of two surfaces. This is the kind of help computers are expected to give. When computing triple integrals, the ultimate challenge is visualization of the region of integration. Since I have no direct experience using this code with students, it is probably inappropriate to describe it here. I would be happy to hear from any readers who might be moved to explore the use of this code in a classroom setting.